Springer Proceedings in Mathematics & Statistics

Volume 21

For further volumes:
http://www.springer.com/series/10533

Springer Proceedings in Mathematics & Statistics

This book series will feature volumes of selected contributions from workshops and conferences in all areas of current research activity in mathematics and statistics, operations research and optimization. In addition to an overall evaluation of the interest, scientific quality, and timeliness of each proposal at the hands of the publisher, every individual contribution is refereed to standards comparable to those of leading journals in the field. This expanded series thus proposes to the research community well-edited and authoritative reports on newest developments in the most interesting and promising areas of mathematical and statistical research today.

Tamás Terlaky • Frank E. Curtis
Editors

Modeling and Optimization: Theory and Applications

Selected Contributions from the MOPTA 2010 Conference

 Springer

Editors
Tamás Terlaky
Department of Industrial and Systems
  Engineering
Lehigh University
Bethlehem, Pennsylvania
USA

Frank E. Curtis
Department of Industrial and Systems
  Engineering
Lehigh University
Bethlehem, Pennsylvania
USA

ISSN 2194-1009   ISSN 2194-1017 (electronic)
ISBN 978-1-4899-9606-0   ISBN 978-1-4614-3924-0 (eBook)
DOI 10.1007/978-1-4614-3924-0
Springer New York Heidelberg Dordrecht London

Mathematics Subject Classification (2010): 49-06, 49Mxx, 65Kxx, 90-06, 90Bxx, 90Cxx

Printed on acid-free paper

Springer is part of Springer Science+Business Media (www.springer.com)

Preface

This volume contains a selection of papers that were presented at the *Modeling and Optimization: Theory and Applications* (MOPTA) conference held at Lehigh University in Bethlehem, Pennsylvania, USA, on August 18–20, 2010. MOPTA 2010 aimed to bring together a diverse group of researchers and practitioners, working on both theoretical and practical aspects of continuous or discrete optimization. The goal was to host presentations on the exciting developments in different areas, and at the same time provide a setting for close interaction among the participants.

The topics covered at MOPTA 2010 varied from algorithms for solving convex, network, mixed-integer, nonlinear, and global optimization problems, and addressed the application of optimization techniques in finance, logistics, health, and other important fields. The five papers contained in this volume represent a sample of these topics and applications, and illustrate the broad diversity of ideas discussed at the conference.

The first part of the name MOPTA highlights the role that modeling plays in the solution of an optimization problem, and indeed, the first two papers in this volume illustrate the benefits of effective modeling techniques. The paper by Mitchell, Pang, and Yu proposes a variety of ways in which a model of a mathematical program with complementarity constraints can be improved with additional constraints. These constraints yield tighter bounds, which can be exploited within certain algorithms. The paper by Chiang and Chen considers new modeling techniques for combinatorial problems arising in communication networks, where it is paramount that the optimization algorithm is distributed.

The next three papers in the volume address the other foci of MOPTA, namely optimization algorithms, theory, and applications. The paper by Iyengar, Phillips, and Stein proposes a first-order method for solving packing–covering semidefinite programs, which arise in important combinatorial problems. Their method is shown to produce highly accurate solutions in a running time faster than interior-point methods. The paper by Howe revives previously unpublished results related to the Shapley–Folkman theorem and extensions of it. Finally, the paper by Romanko, Ghaffari-Hadigheh, and Terlaky expounds on the relationships between multiobjective and parametric optimization, and provides a methodology for the constructing the Pareto efficient frontier. This latter contribution is significant as the frontier can be constructed without discretizing the objective space, thus significantly reducing the computation time.

We thank the sponsors of MOPTA 2010, namely AIMMS, FICO, Gurobi, IBM, Mosek, and SAS. We also thank the host, Lehigh University, as well as the rest of the organizing committee: Pietro Belotti, Imre Pólik, Ted Ralphs, Larry Snyder, Robert Storer, and Aurélie Thiele.

Bethlehem, PA   Frank E. Curtis and Tamás Terlaky

Contents

Chapter 1
Obtaining Tighter Relaxations of Mathematical Programs with Complementarity Constraints

John E. Mitchell, Jong-Shi Pang, and Bin Yu

1.1 Introduction

Mathematical programs with complementarity constraints (MPCCs) arise in many settings. For example, Hobbs et al. [25] discuss applications in deregulated electricity markets; Pang et al. [38] discuss an application in maximum-likelihood-based target classification. The paper [37] shows how the MPCC provides a unifying framework for various modeling paradigms, including hierarchical and inverse optimization. Most recently, the MPCC is used as a tractable formulation for the estimation of pure characteristics models based on empirical market shares [39]. There has been a great deal of research on finding stationary points for MPCCs; see [37] for a list of references. In order to determine a globally optimal solution to an MPCC, it is necessary to find a lower bound on the problem, typically by relaxing the problem. Tightening the relaxation can lead to improved lower bounds, which can be exploited in, for example, branching and domain decomposition schemes. In this chapter, we describe several methods for tightening relaxations of MPCCs. We focus on linear programs with complementarity constraints (LPCCs), a rich subclass of MPCCs. We have previously described logical Benders decomposition

J.E. Mitchell (✉)
Department of Mathematical Sciences, Rensselaer Polytechnic Institute,
Troy, NY 12180-1590, USA
e-mail: mitchj@rpi.edu

J.-S. Pang
Department of Industrial and Enterprise Systems Engineering, University of Illinois,
Urbana, IL 61801, USA
e-mail: jspang@illinois.edu

B. Yu
Department of Industrial and Systems Engineering, Rensselaer Polytechnic Institute, Troy,
NY 12180-1590, USA
e-mail: binyu610@gmail.com

T. Terlaky and F.E. Curtis (eds.), *Modeling and Optimization: Theory and Applications*, Springer Proceedings in Mathematics & Statistics 21, DOI 10.1007/978-1-4614-3924-0_1, © Springer Science+Business Media, LLC 2012

and branch and cut methods for finding globally optimal solutions to LPCCs [26–28]; these methods can be improved by the techniques presented in this chapter. The proposed tightening techniques are expected to be particularly useful for solving the class of convex programs with complementarity constraints, which is a subclass of MPCCs broader than the LPCCs; this extension is presently being investigated.

An LPCC is a linear program with additional complementarity constraints on certain pairs of variables. Because of the complementarity constraints, it is a non-convex, nonlinear disjunctive program. These problems arise in many settings, with the complementarity constraints often used to model logical relations. For example, LPCCs can be used to model bilevel programs, inverse problems, quantile problems, indefinite quadratic programs, and piecewise linear programs; see [28] for a recent summary of such applications.

Given vectors and matrices: $c \in \mathbb{R}^n$, $d \in \mathbb{R}^m$, $e \in \mathbb{R}^m$, $b \in \mathbb{R}^k$, $A \in \mathbb{R}^{k \times n}$, $B \in \mathbb{R}^{k \times m}$, and $C \in \mathbb{R}^{k \times m}$, the LPCC is to find a triple $v := (x, y, w) \in \mathbb{R}^n \times \mathbb{R}^m \times \mathbb{R}^m$ in order to globally solve the optimization problem

$$\Phi \triangleq \underset{x,y,w}{\text{minimize}} \ c^T x + d^T y + e^T w$$

$$\text{subject to} \quad Ax + By + Cw \geq b \tag{1.1}$$

$$0 \leq \quad y \perp \quad w \geq 0$$

where the $\perp$ notation denotes the perpendicularity between two vectors, which in this context pertains to the complementarity of these vectors. Thus, without the orthogonality condition: $y \perp w$, the LPCC is a linear program (LP). With this condition, the LPCC is equivalent to 2^m LPs. The variables x are sometimes called design variables.

Relaxing the complementarity condition leads to a linear programming problem. For some problems, this relaxation can be quite weak, so in this chapter we consider methods for improving the relaxation. Bounds on the variables y and w can be used to construct linear constraints, as we show in Sect. 1.3. These cuts are well known, and we test refinements of the cuts. Disjunctive cuts were developed in the 1970s [7] initially for integer programming. They have been studied extensively and can be generated to cutoff points that violate the complementarity constraints, using the optimal simplex tableau to the LP relaxation and in other ways. We discuss the specialization of disjunctive cuts to LPCCs in Sect. 1.4. The nonconvex quadratic constraint $y^T w \leq 0$ is valid for (1.1); we consider novel convex quadratic relaxations of this constraint in Sect. 1.5. Constraints that are valid on part of the feasible region can be lifted to give constraints valid throughout the feasible region, a technique that can also be used to strengthen other constraints; we discuss lifting in Sect. 1.6. The products of variables can be linearized and a semidefinite constraint imposed to tighten the linearization, as discussed in Sect. 1.7. Each of the families of cuts can be strengthened by exploiting the other families. Thus, the overall strength of the relaxation depends on the order in which the cuts are derived, and it can be further strengthened by repeatedly generating constraints. Under certain conditions, the work of Balas [7] and of Kojima and Tunçel [30] shows that repeated generation of cuts leads to the convex hull of the feasible region of (1.1).

1.2 Problem Generation and Computational Setup

We experimented with a randomly generated collection of LPCCs. The parameters in (1.1) were generated as follows. The matrices A, B, and C are written as $A = [\bar{A}^T, -N^T, I]^T$, $B = [\bar{B}^T, -M^T, 0]^T$, and $C = [0, I, 0]^T$ where 0 denotes a matrix of zeroes of the appropriate dimension, $\bar{A}$ and $\bar{B}$ have $\bar{k} = k - m - n$ rows, and M and N have m rows. Similarly, the right-hand side is written $b = [\bar{b}^T, q^T, 0]^T$ with $\bar{b} \in \mathbb{R}^{\bar{k}}$ and $q \in \mathbb{R}^m$. We set $e = 0$, so the problem is equivalent to

$$\underset{x,y,w}{\text{minimize}}\ c^T x + d^T y$$

$$\text{subject to}\ \ \bar{A}x + \bar{B}y \geq \bar{b}$$

$$x \qquad\qquad \geq 0$$

$$0 \leq y \perp q + Nx + My \geq 0,$$

a standard form in the LPCC literature [28].

The entries in c and d are uniformly distributed integers between 0 and 9, which ensures the problem is not unbounded. The entries in $\bar{A}$, $\bar{B}$, and N are uniformly generated integers between -5 and 5, with a proportion of the entries zeroed out. The matrix $\frac{1}{2}(M + M^T)$ is set equal to LL^T where L is an $m \times r$ matrix whose entries are uniformly generated integers between -5 and 5, with a proportion of the entries zeroed out. This construction ensures that $\frac{1}{2}(M + M^T)$ is positive semidefinite, which is necessary for the approach we develop in Sect. 1.5 and which occurs in some classes of practical instances [28]. The matrix M is then obtained from $\frac{1}{2}(M + M^T)$ by adjusting the nonzero off-diagonal entries by a uniformly distributed random integer between -2 and 2.

To ensure feasibility of (1.1), a solution $\bar{x}$, $\bar{y}$ is generated. The entries in $\bar{x}$ are integers uniformly distributed between 0 and 9. Two-thirds of the entries in $\bar{y}$ are set equal to zero, and the remainder are integers uniformly distributed between 0 and 9. The entries in the right-hand side $\bar{b}$ are chosen so that each slack with the generated solution is an integer uniformly distributed between 1 and 10, so the constraints are strictly satisfied by the generated solution. The third of the entries in q corresponding to the positive components of $\bar{y}$ are chosen so that complementarity is satisfied. Another third of the entries in q are chosen so that the corresponding components of $q + N\bar{x} + M\bar{y}$ are zero, so $\bar{y}_i = \bar{w}_i = 0$ for these entries. The final third of the entries of q are chosen so that the corresponding slack in $q + N\bar{x} + M\bar{y} \geq 0$ is an integer uniformly distributed between 1 and 10. The construction is designed so that it is unlikely that the generated solution $\bar{x}$, $\bar{y}$ is optimal.

The tests in Sects. 1.3–1.5 were run on a single core of an AMD Phenom II X4 955@3.2 GHZ with 4 GB memory, using C++ with callable CPLEX 11.0. All times are reported in seconds. Problems with $m = 100$, 150, and 200 complementarities were solved, with $n = 2$ and $\bar{k} = 20$. The matrices $\bar{A}$, $\bar{B}$, N, and L were either 20% or 70% dense. The rank r of L was either 30 or 60 for $m = 100$, either 30 or 100 for $m = 150$, and either 30 or 120 for $m = 200$. Five problems were solved for each choice of m, sparsity, and rank, leading to a total of 60 problems.

1.3 Linear Constraints Based on Bounds on the Variables

1.3.1 Construction of the Constraints

Given finite upper bounds y_i^u and w_i^u on y_i and w_i, respectively, the constraint

$$w_i^u y_i + y_i^u w_i \leq y_i^u w_i^u, \tag{1.2}$$

which we term a *bound cut*, is valid for the LPCC (1.1), because of the complementarity restriction on y_i and w_i. The bounds y_i^u and w_i^u may not be readily available and can be calculated by solving linear programming problems. Before calculating bounds, it is useful to find a good feasible solution to the LPCC, using either a heuristic or a nonlinear programming solver such as KNITRO [20] or FILTER [23]. The value of this solution provides an upper bound Φ^{UB} on the optimal value Φ of (1.1) and so the constraint

$$c^T x + d^T y + e^T w \leq \Phi^{UB} \tag{1.3}$$

is valid for any optimal solution to (1.1). We let S denote the set of feasible solutions to (1.1) that satisfy (1.3). The convex hull of S is a polyhedron and it can be outer-approximated by

$$\Xi \triangleq \{(x, y, w) \in \mathbb{R}^{n+2m} : Ax + By + Cz \geq b, c^T x + d^T y + e^T w \leq \Phi^{UB}, y \geq 0, w \geq 0\}. \tag{1.4}$$

If an inequality description of the convex hull of S was known, then the LPCC could be determined by solving the linear program of minimizing the objective over the convex hull. The aim of constructing the inequalities in this chapter is to obtain a tighter approximation to the convex hull than Ξ. Under certain conditions, this constraint defines a facet of the convex hull of feasible solutions to the LPCC, as shown by De Farias et al. [22] for the case of a knapsack LPCC. It is straightforward to prove the following proposition.

Proposition 1.1. *Assume either (i) $y_i \geq 0$ defines a facet of S and $w_i \leq w_i^u$ defines a facet of $\{(x,y,s) \in S : y_i = 0\}$ and there is a point in S with $y_i = y_i^u$ or (ii) $w_i \geq 0$ defines a facet of S and $y_i \leq y_i^u$ defines a facet of $\{(x,y,s) \in S : w_i = 0\}$ and there is a point in S with $w_i = w_i^u$. Then (1.2) defines a facet of S.*

Proof. We prove case (i). If $y_i^u = 0$, then the result is immediate. If $y_i^u > 0$, then the facet assumptions together with a point satisfying the constraint (1.2) at equality that is not in $\{(x,y,s) \in S : y_i = 0\}$ leads to the conclusion, from standard lifting arguments. $\quad\square$

We will give a generalization of this proposition later, in Proposition 1.3. In principle, an upper bound on y_i can be found by solving the linear program

$$\underset{x,y,w}{\text{maximize}} \qquad y_i$$

$$\text{subject to} \quad Ax + By + Cw \geq b$$

$$c^T x + d^T y + e^T w \leq \Phi^{\text{UB}} \qquad (1.5)$$

$$w_i = 0$$

$$0 \leq \quad y, \qquad w \geq 0,$$

and an upper bound on w_i can be constructed similarly. These bounds may be infinite; they can be tightened by further exploiting complementarity. For example, we experimented with the following procedure, choosing p equal to 2 or 3:

Bound tightening procedure

Step 0. Find initial upper bound y_i^u by solving (1.5). Let $(\hat{x}, \hat{y}, \hat{w})$ denote the optimal solution.

Step 1. Let $s_j = \hat{y}_j \hat{w}_j$ for $j = 1, \ldots, m$. Let J denote the indices of the largest p components of s.

Step 2. Solve the 2^p linear programs of the form (1.5) with the additional constraints that for each $j \in J$ either $y_j = 0$ or $w_j = 0$. Update y_i^u to be the largest of the optimal values of these 2^p linear programs.

A similar procedure is used to improve w_i^u.

It may be advisable computationally to limit the number of bounds calculated. One approach to do this is to first solve the LP relaxation of (1.1) and then only calculate bounds for variables where the complementarity constraint is violated. Similarly, the bound-tightening procedure could be used only for constraints (1.2) that are tight at the solution to the LP relaxation of (1.1).

Table 1.1 Bound cuts for LPCCs

# Splits	# Refine	% Gap closed			CPU time (s)			Sufficient
		100	150	200	100	150	200	
0	0	45.5	48.0	51.2	1.4	6.8	22.5	0
	4	69.8	75.9	74.9	16.3	76.5	257.7	10
	8	72.7	76.3	75.3	30.8	151.4	482.9	15
1	0	55.8	55.4	57.4	3.9	18.3	60.2	1
	4	76.9	84.1	76.6	42.1	194.3	660.5	15
	8	81.1	85.5	77.1	73.4	367.0	1,171.1	17
2	0	58.3	59.0	60.0	6.0	29.0	93.1	1
	4	82.6	86.1	77.6	61.8	285.6	999.7	16
	8	86.6	87.8	79.5	104.7	542.2	1,786.1	19
3	0	60.6	60.1	61.8	10.4	49.9	159.4	1
	4	85.6	88.9	80.0	102.7	505.1	1,672.4	18
	8	91.3	90.6	80.6	171.6	901.6	3,150.6	21

If upper bounds are available for all the variables y_i and w_i, then (1.1) is equivalent to the following integer programming problem:

$$\begin{aligned} \underset{x,y,w}{\text{minimize}} \quad & c^T x + d^T y + e^T w \\ \text{subject to} \quad & Ax + By + Cw \geq b \\ & 0 \leq \quad y \leq Y^u z \\ & 0 \leq \quad w \leq W^u(\mathbf{1} - z) \\ & z \in \{0,1\}^m \end{aligned} \qquad (1.6)$$

Here, $\mathbf{1}$ denotes the vector of all ones, and Y^u and W^u are diagonal matrices, with diagonal entries equal to the bounds y^u and w^u.

1.3.2 Computational Results

An experiment was performed to test the ideas presented so far. The results were not significantly affected by sparsity or rank, so we aggregate into 20 problems each with 100, 150, or 200 complementarities. The results are contained in Table 1.1. The first column reports the choice of p in the bound-tightening procedure. The second column gives the number of refinements of the bound-tightening procedure: once bounds have been found, the bound finding LP (1.5) can be tightened. The third, fourth, and fifth columns give the relative improvement in the gap between the lower bound on (1.1) given by its LP relaxation and the optimal value of (1.1); each entry in these columns is averaged over 20 problems. The sixth, seventh, and eighth

Table 1.2 Solving to optimality with bound cuts for LPCCs

# Splits	# Refine	Successful			Solve time (s)			Total time (s)		
		100	150	200	100	150	200	100	150	200
No cuts		20	11	5	98	3,862	5,852	98	3,862	5,852
0	0	20	19	12	13	632	3,265	**14**	639	3,290
	4	20	20	18	6	102	1,558	22	**179**	**1,816**
	8	20	20	18	5	66	1,382	36	**217**	**1,871**
1	0	20	19	12	12	843	2,992	**16**	861	3,052
	4	20	20	19	4	64	906	46	**258**	**1,567**
	8	20	20	19	3	78	942	76	445	**2,113**
2	0	20	20	13	10	496	2,962	**16**	525	3,055
	4	20	20	19	4	67	930	66	811	**1,930**
	8	20	20	19	2	60	977	107	602	2,763
3	0	20	19	13	8	536	2,763	23	586	2,922
	4	20	20	19	3	60	795	106	565	2,467
	8	20	20	19	2	52	841	174	954	3,992

columns give the run time, with each entry averaged over 20 instances. The final column notes the number of problems (out of 60) for which the bound cuts were sufficient to prove global optimality.

The bound cuts are effective at closing a large proportion of the duality gap. However, they are quite expensive, especially with additional refinements and splits. It is noticeable that the smaller problems benefit slightly more than the larger problems from additional refinements and splits. The bound cuts are surprisingly effective at proving optimality for these problems, with over one-third of the problems solved to optimality with the most extensive version of the cuts. Refining the bound cuts eight times closes between 47% and 78% of the gap remaining after the addition of just the initial bound cuts.

We also used CPLEX 11 to try to solve these problems to optimality, on the same computer hardware. The callable library version of CPLEX was used, which allows the representation of disjunctive constraints using indicator constraints, so it is possible to work directly with formulation (1.1) together with constraints (1.2) and (1.3). An initial feasible solution was found with a heuristic. A time limit of 2 h was placed on each run. The results are contained in Table 1.2.

The columns indicate the number of problems that were solved in the 2 h time limit, and the average time for each set of 20 instances. Both the time to solve the problems to optimality after adding the cuts and the total time including the cut-generation time are included in the table. All total times that are within 50% of the minimum are highlighted. The cuts lead to dramatic improvements in the ability of CPLEX to solve the instances and in the total time required. It is not worthwhile to refine the cuts for the smaller problems with $m = 100$ complementarities. For larger problems, the time invested in generating the cuts and refining them can lead to strong overall performance, with one good option being to use four refinements along with one split. Additional refinements or splits aid the algorithm in finding a solution within the 2 h limit; only one problem cannot be solved within 2 h with the more extensive cut generation choices. Even better performance could probably be obtained by generating and adding the bound cuts selectively, based on the solution to the LP relaxation; in this chapter, we are examining the strength of the class of cuts as a whole.

1.4   Linear Constraints Based on Disjunctive Programming

1.4.1   Disjunctive Cuts

Valid constraints can be constructed from any point in Ξ that is not in the convex hull of feasible solutions to (1.1), using a disjunctive programming approach. If $y_i w_i > 0$ in an extreme point optimal solution to a relaxation of (1.1) then it is not in the convex hull, so valid constraints are constructed that are satisfied by all points in Ξ with $y_i = 0$ and by all points in Ξ with $w_i = 0$. A cut generation linear program can be formulated to find such valid constraints. Balas [7, 8] developed many of the results regarding disjunctive cuts for integer programming. Many of the approaches used in integer programming are also useful in more general disjunctive programs. For example, Audet et al. [4] consider disjunctive cuts for bilevel linear programs. For good recent surveys of methods of generating disjunctive cuts see [11, 16, 40, 44]. It was shown empirically in the 1990s that general cuts such as disjunctive cuts [9] and Gomory cuts [10] could be very effective for general integer programs. Theoretically, the convex hull of an LPCC can be obtained using the lift-and-project procedure, since the disjunctions are facial [21]. Also of interest is recent work showing that disjunctive cuts can be effective for mixed integer nonlinear programming problems [14, 45, 46, 49]. Judice et al. [29] investigated disjunctive cuts for problems with complementarity constraints.

Let $v = (x, y, w) \in \mathbb{R}^{n+2m}$ and let $\hat{v}$ be the optimal solution to the LP relaxation. A general disjunctive cut for the union of a family of polyhedra is an inequality that is valid for each polyhedron in the family. It can be obtained by solving a cut generation LP which ensures that the cut is dominated by a nonnegative linear combination of the valid constraints for each polyhedron. This cut generation LP can be large, so methods have been developed to find cuts without solving the full cut generation LP. The optimal simplex tableau for the linear programming relaxation can be used directly to generate constraints that cut off $\hat{v}$ if it violates the complementarity restrictions. In particular, if $y_i w_i > 0$, then the two rows of the simplex tableau corresponding to the basic variables y_i and w_i can be written as follows, where R denotes the set of nonbasic variables:

$$y_i \quad + \sum_{j \in R} \hat{a}_j^{y_i} v_j = \hat{y}_i$$

$$w_i + \sum_{j \in R} \hat{a}_j^{w_i} v_j = \hat{w}_i. \tag{1.7}$$

The disjunction $y_i = 0 \vee w_i = 0$ is equivalent to the disjunction

$$\sum_{j \in R} \frac{\hat{a}_j^{yi}}{\hat{y}_i} v_j \geq 1 \quad \vee \quad \sum_{j \in R} \frac{\hat{a}_j^{wi}}{\hat{w}_i} v_j \geq 1$$

since y_i and w_i are nonnegative variables. Let $\alpha_j = \max \left\{ \frac{\hat{a}_j^{yi}}{\hat{y}_i}, \frac{\hat{a}_j^{wi}}{\hat{w}_i} \right\}$ for $j \in R$. We can construct the following valid constraint for (1.1):

$$\sum_{j \in R} \alpha_j v_j \geq 1. \tag{1.8}$$

This constraint is violated by $\hat{v}$ since $\hat{v}_j = 0$ for $j \in R$. It is valid because either $y_i = 0$ or $w_i = 0$ in any feasible solution, so the sum of the nonbasic variables in (1.7) must be equal to the right-hand side for at least one of the constraints, and the sum of the nonbasic variables (scaled by the right-hand side) is overestimated by the sum given in (1.8). This is called a *simple cut* by Audet et al. [5], and is based on intersection cuts for 0–1 programming [6] and has also been investigated by Balas and Perregaard [12].

If the complementarity restrictions for components i and k are both violated by $\hat{v}$, then the corresponding four rows of the simplex tableau can be combined to obtain valid constraints for

$$\Xi^{ik} \triangleq \Xi \cap \{v \mid y_i w_i = 0\} \cap \{v \mid y_k w_k = 0\}.$$

In particular, we can set up the following cut generation LP which generates a constraint that is valid for each of the four pieces of Ξ^{ik} corresponding to each assignment of the i and k complementarity relationships. Any feasible solution to this LP gives a valid constraint of the form (1.8) that cuts off $\hat{v}$:

$$\begin{aligned}
\underset{\alpha,u}{\text{minimize}} \quad & \sum_{j \in R} \alpha_j \\
\text{subject to} \quad & \alpha_j \geq u_{1i} \frac{\hat{a}_j^{yi}}{\hat{y}_i} + u_{1k} \frac{\hat{a}_j^{yk}}{\hat{y}_k} \quad \forall j \in R \\
& \alpha_j \geq u_{2i} \frac{\hat{a}_j^{yi}}{\hat{y}_i} + u_{2k} \frac{\hat{a}_j^{wk}}{\hat{w}_k} \quad \forall j \in R \\
& \alpha_j \geq u_{3i} \frac{\hat{a}_j^{wi}}{\hat{w}_i} + u_{3k} \frac{\hat{a}_j^{yk}}{\hat{y}_k} \quad \forall j \in R \\
& \alpha_j \geq u_{4i} \frac{\hat{a}_j^{wi}}{\hat{w}_i} + u_{4k} \frac{\hat{a}_j^{wk}}{\hat{w}_k} \quad \forall j \in R \\
& 1 \leq u_{pi} + u_{pk} \qquad \text{for } p = 1, \dots, 4 \\
& u_{pq} \geq 0 \qquad \text{for } p = 1, \dots, 4, \, q = i, k.
\end{aligned} \tag{1.9}$$

Table 1.3 Computational results with disjunctive cuts

Cut type	# Refine	% Gap closed			CPU time (s)			Sufficient
		100	150	200	100	150	200	
Disjunctive	$m/8$	65.3	65.7	61.9	18.4	162.9	708.5	0
	$m/2$	75.9	75.1	70.8	78.6	605.2	2,547.6	1
Simple	$m/8$	32.7	25.3	21.4	0.1	0.8	3.4	0
	$m/2$	34.6	26.0	21.9	2.6	23.8	118.1	0

The first four constraints correspond to different pieces of Ξ^{ik} and ensure that constraint (1.8) is dominated by a nonnegative combination of the constraints for that piece. For example, the first constraint corresponds to the piece with $y_i = y_k = 0$, and ensures the constraint is dominated by a combination of the corresponding nonbasic parts of (1.7). The objective function together with the constraints $1 \leq u_{pi} + u_{pk}$ act to normalize the constraint generation LP; other normalizations could be used instead. This linear program is far smaller than the standard disjunctive cut generation LP; it in effect constrains many variables from the standard LP to be equal to 0. Balas and Perregaard [11] discuss similar methods for making the cut generation LP easier to solve. It should be noted that the standard simple cut is an optimal solution to a constrained version of (1.9), obtained by adding the constraints $u_{pi} = 1$ and $u_{pk} = 0$ for $p = 1, \dots, 4$.

1.4.2 Computational Results

The disjunctive cuts and simple cuts were tested for the same problems as in Sect. 1.2, in the same computational environment. Computational results are contained in Table 1.3. The cuts were refined successively, with the number of refinements proportional to the number of complementarities and given in the second column of the table. Columns 3–9 of the table have the same meanings as in Table 1.1.

The general disjunctive cuts are far more effective than the simple cuts, but they are considerably more expensive. Additional refinement is quite useful for the disjunctive cuts, but far less so for the simple cuts. Audet et al. [5] have experimented with disjunctive cuts for LPCCs arising from bilevel programs, with encouraging results.

1.5   Convex Quadratic Constraints

1.5.1   Construction of the Constraints

The complementarity constraint

$$0 \leq y \perp w \geq 0$$

is equivalent to the nonnegativity constraints $y, w \geq 0$ together with the nonconvex quadratic constraint

$$y^T w \leq 0. \tag{1.10}$$

In this section, we consider convex quadratic relaxations of (1.10). We assume that w can be written as a linear function of x and y, so

$$w = q + Nx + My \tag{1.11}$$

where q, N, and M are dimensioned appropriately. We express the complementarity restriction in terms of x and y, so the number of constraints depends on the dimension n of x rather than on the number m of complementarity constraints. We have

$$y^T w = q^T y + y^T Nx + \tfrac{1}{2} y^T \tilde{M} y$$

where $\tilde{M} = M + M^T$, so we look for convex relaxations of the quadratic constraint

$$q^T y + y^T Nx + \tfrac{1}{2} y^T \tilde{M} y \leq 0. \tag{1.12}$$

Let p denote the number of nonnegative eigenvalues of $\tilde{M}$ and construct an eigen-decomposition of $\tilde{M}$ as

$$\tilde{M} = V \Lambda V^T$$

where V is an orthogonal matrix with columns denoted v_i, Λ is a diagonal matrix, and the diagonal entries λ_i of Λ are arranged in decreasing order. Let k denote the rank of N and construct a factorization $N = \Gamma^T \Psi$, where Γ is a $k \times m$ matrix and Ψ is a $k \times n$ matrix. With the definition of k-dimensional variables $\tilde{y}$ and $\tilde{x}$, and the addition of the constraints

$$\tilde{y} = \Gamma y \tag{1.13}$$

$$\tilde{x} = \Psi x, \tag{1.14}$$

constraint (1.12) is equivalent to the constraint

$$q^T y + \sum_{j=1}^{k} \tilde{y}_j \tilde{x}_j + \tfrac{1}{2} \sum_{i=1}^{p} \lambda_i (v_i^T y)^2 \leq \tfrac{1}{2} \sum_{i=p+1}^{m} |\lambda_i| (v_i^T y)^2$$

or equivalently the convex quadratic constraint

$$q^T y + \sum_{j=1}^{k} \sigma_k + \tfrac{1}{2} \sum_{i=1}^{p} \lambda_i (v_i^T y)^2 \leq \tfrac{1}{2} \sum_{i=p+1}^{m} |\lambda_i| \pi_i \tag{1.15}$$

with the additional nonconvex constraints

$$\tilde{y}_j \tilde{x}_j = \sigma_j \qquad j = 1, \ldots, k \tag{1.16}$$

$$(v_i^T y)^2 \geq \pi_i \qquad i = p+1, \ldots, m \tag{1.17}$$

When the feasible region for $\tilde{y}_j$ and $\tilde{x}_j$ is given by a rectangle, it was shown by Al-Khayyal and Falk [2] that the lower convex envelope and upper concave envelope of (1.16) are given by the following *McCormick cuts* [34]:

$$
\begin{aligned}
\tilde{x}_j^l \tilde{y}_j + \tilde{y}_j^l \tilde{x}_j &\leq \sigma_j + \tilde{x}_j^l \tilde{y}_j^l & j &= 1, \ldots, k \\
\tilde{x}_j^u \tilde{y}_j + \tilde{y}_j^u \tilde{x}_j &\leq \sigma_j + \tilde{x}_j^u \tilde{y}_j^u & j &= 1, \ldots, k \\
\tilde{x}_j^l \tilde{y}_j + \tilde{y}_j^u \tilde{x}_j &\geq \sigma_j + \tilde{x}_j^l \tilde{y}_j^u & j &= 1, \ldots, k \\
\tilde{x}_j^u \tilde{y}_j + \tilde{y}_j^l \tilde{x}_j &\geq \sigma_j + \tilde{x}_j^u \tilde{y}_j^l & j &= 1, \ldots, k,
\end{aligned}
\tag{1.18}
$$

where $\tilde{y}_j^l$, $\tilde{y}_j^u$, $\tilde{x}_j^l$, and $\tilde{x}_j^u$ denote the bounds on $\tilde{y}_j$ and $\tilde{x}_j$. These constraints are exploited in packages for nonconvex optimization, including BARON [43, 50], αBB [1], and COUENNE [15]. Tightenings of these inequalities combining together terms for several indices j have been recently investigated by Bao et al. [13].

We look for convex quadratic relaxations of (1.16) and linear relaxations of (1.17) that exploit the structure of the other linear constraints on $\tilde{y}_j$ and $\tilde{x}_j$. If $v_i^l \leq v_i^T y \leq v_i^u$ for all valid choices of y, then the concave envelope of (1.17) is

$$\pi_i \leq (v_i^l + v_i^u) v_i^T y - v_i^l v_i^u. \tag{1.19}$$

For any scalar $\alpha > 0$, (1.16) is equivalent to the constraint

$$\frac{1}{4\alpha} (\tilde{y}_j + \alpha \tilde{x}_j)^2 \leq \sigma_j + \frac{1}{4\alpha} (\tilde{y}_j - \alpha \tilde{x}_j)^2$$

which can be relaxed to the convex quadratic constraint

$$\frac{1}{4\alpha} (\tilde{y}_j + \alpha \tilde{x}_j)^2 \leq \sigma_j + \frac{1}{4\alpha} ((\alpha^l + \alpha^u)(\tilde{y}_j - \alpha \tilde{x}_j) - \alpha^l \alpha^u) \tag{1.20}$$

where α^l and α^u denote lower and upper bounds respectively on $\tilde{y}_j - \alpha \tilde{x}_j$. Methods for choosing α are discussed in [35]. It is also shown in this reference that (1.20) can define part of the envelope of (1.16). Further, for certain configurations of the feasible $(\tilde{y}_j, \tilde{x}_j)$ region, inequalities (1.20) together with (1.18) define the lower convex envelope and upper concave envelope of (1.16). Thus, we propose to relax (1.12) using the linear constraints (1.13), (1.14), (1.18), and (1.19), and the convex quadratic constraints (1.15) and (1.20). We have the following proposition regarding the strength of (1.20).

Proposition 1.2 ([35]). *Let $\mathcal{P}^j$ denote the projection of a polyhedral relaxation of (1.1) onto the $(\tilde{x}_j, \tilde{y}_j)$ plane. Let*

$$\bar{\mathcal{P}}^j \triangleq \{(\tilde{x}_j, \tilde{y}_j, \sigma_j) \mid (\tilde{x}_j, \tilde{y}_j) \in \mathcal{P}^j,\, \sigma_j = \tilde{x}_j \tilde{y}_j\}.$$

(a) Assume $\mathcal{P}^j$ has the form

$$\tilde{x}_j^L \leq \tilde{x}_j \leq \tilde{x}_j^U,\ \tilde{y}_j^L \leq \tilde{y}_j \leq \tilde{y}_j^U,\ L^j \leq \tilde{x}_j - \bar{\alpha}\tilde{y}_j \leq U^j$$

for parameters $\bar{\alpha}$, $\tilde{x}_j^L$, $\tilde{x}_j^U$, $\tilde{y}_j^L$, $\tilde{y}_j^U$, L^j and U^j.

1. *If $\bar{\alpha} > 0$, $\tilde{x}_j^U - \tilde{x}_j^L = \bar{\alpha}(\tilde{y}_j^U - \tilde{y}_j^L)$, and $L^j + U^j = \tilde{x}_j^L + \tilde{x}_j^U - \bar{\alpha}(\tilde{y}_j^L + \tilde{y}_j^U)$ then the lower convex underestimator of σ_j over $\mathcal{P}^j$ is given by (1.18) together with (1.20) with $\alpha = \bar{\alpha}$.*
2. *If $\bar{\alpha} < 0$, $\tilde{x}_j^U - \tilde{x}_j^L = -\bar{\alpha}(\tilde{y}_j^U - \tilde{y}_j^L)$, and $L^j + U^j = \tilde{x}_j^L + \tilde{x}_j^U - \bar{\alpha}(\tilde{y}_j^L + \tilde{y}_j^U)$ then the upper concave overestimator of σ_j over $\mathcal{P}^j$ is given by (1.18) together with (1.20) with $\alpha = \bar{\alpha}$.*

(b) Let $\mathcal{P}^j$ have the form

$$L_1^j \leq \tilde{x}_j - \bar{\alpha}\tilde{y}_j \leq U_1^j,\ L_2^j \leq \tilde{x}_j + \bar{\alpha}\tilde{y}_j \leq U_2^j$$

for some $\bar{\alpha} > 0$. If $\mathcal{P}^j$ is nonempty then the convex envelope of $\bar{\mathcal{P}}^j$ is given by (1.20) with $\alpha = \pm\bar{\alpha}$. □

1.5.2 Computational Results

The test problems and computational environment were the same as in Sect 1.2. The convex quadratic program solver CPLEX reported that the matrix $M + M^T$ was not positive semidefinite for 5 of the 60 instances, due to numerical errors, and so these problems were not solved. Consequently, each entry in the "gap" columns and the "time" columns represents a mean of 20, 19, or 16 instances. The performance of refining the McCormick cuts procedure is compared with refining the bound generation procedure in Fig. 1.1.

Using convex relaxations of the constraint $y^T w \leq 0$ is very effective for this class of problems, giving better bounds than from either the bound cuts of Sect. 1.3.2

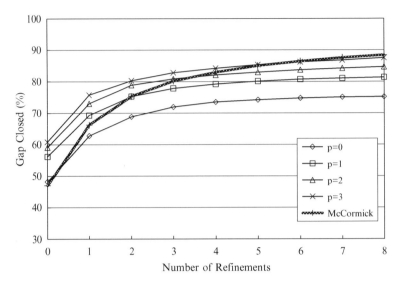

Fig. 1.1 Comparison of average gap closed by refinements of bound cuts and McCormick cuts

Table 1.4 McCormick cuts and convex quadratic cuts

	% Gap closed			CPU time (s)			
	100	150	200	100	150	200	Sufficient
McCormick	44.3	49.1	47.9	0.2	0.5	1.2	0
+8 refines	83.5	91.5	90.7	9.7	29.4	67.6	6
+Quadratic	83.5	91.5	90.8	10.0	30.3	69.3	8

or the disjunctive cuts of Sect. 1.4.2, in far less time. The quadratic constraint
(1.20) is only marginally helpful for these problems; simply iteratively tightening
the McCormick bounds (1.18) works very well.

We also attempted to solve these problems to optimality, using a combination of
bound cuts and McCormick cuts. We used the bound cut procedure with no splits
and four refinements, since these bounds can be found in a moderate amount of time
(Table 1.1) and the cuts are quite effective at solving the problems (Table 1.2). After
adding these refined bound cuts, the McCormick cut generation procedure was used,
with eight refinements, as in Table 1.4. For these runs, CPLEX reported that the
matrix $M + M^T$ was indefinite for 4 of the 60 instances, and it was unable to solve an
additional three instances in the 2 h time limit. This left 20, 19, and 14 instances with
100, 150, and 200 complementarities, respectively. These 53 instances were also all
solved in the 2 h limit using just the bound cuts. The runtimes with just the bound
cuts and also with the bound cuts together with the McCormick cuts are contained in
Table 1.5. Five additional problems were solved at the root node through the addition
of the McCormick cuts. The additional time required to generate the McCormick
cuts is worthwhile for the larger instances.

Table 1.5 Solving to optimality with McCormick cuts

	Solve time (s)			Total time (s)		
	100	150	200	100	150	200
Bound cuts	6	74	626	48	268	1,286
Bound + McCormick	6	37	368	58	260	1,095

When using just the bound cuts, 2 of the 200 complementarity instances cannot be solved in 2 h, and 5 other instances require at least 20 min. The two instances requiring at least 2 h still cannot be solved within that time limit with the addition of McCormick cuts, but the duality gap at the end of the time limit has been noticeably reduced. CPLEX reported that $M + M^T$ was indefinite for one of the remaining hard instances, and it was unable to solve one of them within the 2 h limit. The McCormick cuts reduced the run time on each of the other three hard instances, with the average solve time dropping from 2,681 to 1,598 s.

1.6 Lifting Constraints

Lifting is a methodology for modifying a constraint that is valid on one part of the feasible region so that it is valid throughout the feasible region. Let $\mathcal{P} \subseteq \mathbb{R}^n_+$ denote a polyhedron and let $\mathcal{P}^0 \triangleq \{x \in \mathcal{P} : x_i = 0\}$ for a fixed component i. Given a constraint $a^T x \geq \beta$ that is valid on $\mathcal{P}^0$, a lifting procedure can be used to extend this constraint so that it is valid throughout $\mathcal{P}$. This idea is widely employed in integer programming [36].

De Farias et al. [22] describe a lifting procedure for LPCCs with $k = 1$ and all coefficients negative. They show that (1.2) defines a facet of the feasible region under some mild conditions on the coefficients of the problem. They also show that if an inequality is facet defining when $w_i = 0$, then it can be lifted to a facet-defining inequality for the whole feasible region by solving a parametric linear program, again under certain mild assumptions. Further, they describe various families of facet-defining inequalities. Richard and Tawarmalani [42] generalize lifting to nonlinear programs. Given an affine minorant of a function $f(x,y) : \mathbb{R}^{m+n} \to \mathbb{R}$ that is valid for a particular choice of y, they show how the affine minorant can be extended to be a minorant for all y. Of particular interest is the case when $f(x,y)$ is a membership indicator function for a set S, which is zero for points in the set and infinite otherwise.

The general lifting framework of [42] can be specialized to LPCCs by using a membership indicator function for the set S of feasible points to the LPCC (1.1) that satisfy the objective function bound constraint (1.3). Given disjoint subsets I_1, $I_2 \subseteq \{1, \ldots, m\}$, let

$$S^{I_1, I_2} \triangleq \{v \mid (x, y, w) \in S : y_i = 0 \ \forall i \in I_1, w_i = 0 \ \forall i \in I_2\}.$$

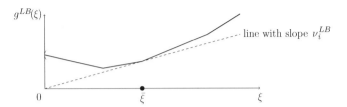

Fig. 1.2 Illustration of the function $g^{\text{LB}}(\xi)$ for $0 < \xi \le y_i^{\text{UB}}$

Let

$$\alpha^T v \le \beta$$

be a valid constraint for S^{I_1,I_2} and let $i \in I_1$. We want to extend the constraint so that it is valid for $S^{I_1 \setminus i, I_2}$, constructing a constraint of the form

$$\alpha^T v + v_i y_i \le \beta$$

for some constant v_i. As shown in [42], it suffices to choose v_i so that $v_i y_i$ underestimates

$$g(\xi) \triangleq \inf\left\{ \beta - \alpha^T v \mid v \in S^{I_1 \setminus i, I_2}, y_i = \xi \right\}.$$

If $S^{I_1 \setminus i, I_2}$ is compact, the parameter v_i can be determined by solving a fractional program:

$$v_i \triangleq \inf\left\{ \frac{\beta - \alpha^T v}{y_i} \mid v \in S^{I_1 \setminus i, I_2}, y_i > 0 \right\},$$

which can be solved as a parametric LPCC. For example, (1.2) can be derived by lifting the inequality $w_i \le w_i^{\text{UB}}$ that is valid on $S^{i,\emptyset}$, when we obtain $v_i = w_i^{\text{UB}}/y_i^{\text{UB}}$. The lifting procedure can be used to obtain facets of conv(S) using the following proposition from [42] (see also [22, 36]) specialized to the case of the LPCC:

Proposition 1.3. *If $\alpha^T v \le \beta$ defines a facet of conv(S^{I_1,I_2}), if the dimension of $S^{I_1 \setminus i, I_2}$ is one more that the dimension of S^{I_1,I_2}, and if the constraint $\alpha^T v + v_i y_i \le \beta$ is valid for $S^{I_1 \setminus i, I_2}$ and satisfied at equality by at least one point in $S^{I_1 \setminus i, I_2} \setminus S^{I_1,I_2}$, then $\alpha^T v + v_i y_i \le \beta$ defines a facet of $S^{I_1 \setminus i, I_2}$.* $\qquad\square$

Determining the optimal choice for v_i is itself a hard problem, so a relaxation can be used in order to obtain a lower bound v_i^{LB}. Any lower bound will provide a constraint

$$\alpha^T v + v_i^{\text{LB}} y_i \le \beta$$

that is valid throughout $S^{I_1 \setminus i, I_2}$. For example, a parametric linear programming problem can be solved to find a lower bound $g^{\text{LB}}(\xi)$. The function $g^{\text{LB}}(\xi)$ is then a piecewise linear convex function in $0 < \xi \le y_i^{\text{UB}}$, as illustrated in Fig. 1.2. In order to construct a lower bound using parametric linear programming, it is necessary to have polyhedral outer approximations $\bar{S}^{l_1 \setminus i, l_2} \supseteq S^{l_1 \setminus i, l_2}$ and $\bar{S}^{l_1, l_2} \supseteq S^{l_1, l_2}$. We have $g^{LB}(0) \geq 0$, since we can assume that the constraint $\alpha^T v \leq \beta$ is included in the description of $\bar{S}^{l_1, l_2}$. The left-hand limit of $g^{LB}(\xi)$ as $\xi \to 0+$ is found by solving the linear program

$$g^{LB}(0+) = \min \{ \beta - \alpha^T v : v \in \bar{S}^{l_1 \setminus i, l_2}, w_i = 0, y_i = 0 \}.$$

If $g^{LB}(0+) < 0$, then it is not possible to lift the constraint using the relaxation, since the resulting bound on v_i is $-\infty$. Using a parametric LP approach, the lower bound v_i^{LB} is chosen to equal

$$v_i^{LB} = \inf \left\{ \frac{\beta - \alpha^T v}{\xi} \mid v \in \bar{S}^{l_1 \setminus i, l_2}, w_i = 0, 0 < \xi \leq y_i^{UB} \right\},$$

illustrated in the figure, with $\bar{\xi}$ equal to the arginf. It is the slope of the greatest homogeneous affine minorant of $g^{LB}(\xi)$, and may well be negative. Note that if $g^{LB}(0+) = 0$, then v_i^{LB} is the slope of the first line segment of $g^{LB}(\xi)$.

Example 1.1. Consider the following LPCC feasible region:

$$2x_1 - y_1 \leq 4 \tag{1.21}$$

$$2x_1 + y_1 \leq 6 \tag{1.22}$$

$$x_1 + 2y_1 \leq 6 \tag{1.23}$$

$$y_1 - y_2 \leq 2 \tag{1.24}$$

$$x_1, x_2 \geq 0 \tag{1.25}$$

$$0 \leq y_1 \perp w_1 \triangleq 3x_1 - 2y_1 + 2 \geq 0 \tag{1.26}$$

$$0 \leq y_2 \perp w_2 \triangleq 3x_1 + x_2 + 6y_1 - 14 \geq 0. \tag{1.27}$$

When $y_1 = 0$, it follows from (1.21) that $x_1 \leq 2$. We lift this constraint so that it is valid when $y_1 > 0$, giving a constraint of the form

$$x_1 + v_1 y_1 \leq 2.$$

By complementarity (1.26), if $y_1 > 0$ then $w_1 = 0$. We can calculate the function $g^{LB}(\xi)$ for $\xi > 0$ using the following LP:

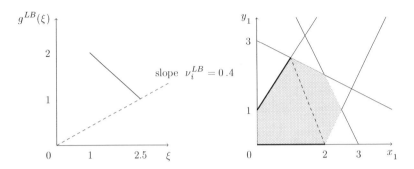

Fig. 1.3 (**a**) Illustration of the function $g^{LB}(\xi)$ for $0 < \xi$ in Example 1.1. (**b**) Projection of example on (x_1, y_1) space, with the LP relaxation feasible region *shaded*. The *thick line* segments indicate points satisfying $y_1 w_1 = 0$. The *dashed line* is the lifted constraint (1.28)

$$g^{LB}(\xi) = \text{minimize } 2 - x_1$$

$$\text{subject to} \quad 2x_1 \quad \leq 4 + \xi$$
$$2x_1 \quad \leq 6 - \xi$$
$$x_1 \quad \leq 6 - 2\xi$$
$$- y_2 \leq 2 - \xi$$
$$3x_1 \quad = 2\xi - 2$$
$$3x_1 + x_2 \geq 14 - 6\xi$$
$$x_1, x_2, y_2 \geq 0.$$

This gives

$$g^{LB}(\xi) = \begin{cases} +\infty & \text{if } 0 < \xi < 1 \\ \dfrac{8 - 2\xi}{3} & \text{if } 1 \leq \xi \leq 2.5 \\ +\infty & \text{if } 2.5 < \xi \end{cases}$$

as illustrated in Fig. 1.3a. The greatest slope for a homogeneous affine minorant of $g^{LB}(\xi)$ is $v_1^{LB} = 0.4$, leading to the lifted constraint

$$x_1 + 0.4y_1 \leq 2. \tag{1.28}$$

The projection of the LP relaxation onto the (x_1, y_1) space is illustrated in Fig. 1.3b. Note that if the complementarity condition in (1.27) is imposed when calculating $g(\xi)$, then either $y_1 \leq 2$ or $3x_1 + 6y_1 \leq 14$. This leads to a slightly larger coefficient $v = 2/3$, resulting in a somewhat stronger lifted constraint. Thus, $g^{LB}(\xi)$ is a strict minorant of $g(\xi)$ for this example. □

A constraint that is valid for S^{I_1,I_2} can be successively lifted in all the variables $y_i, i \in I_1$ and $w_i, i \in I_2$, leading to a valid constraint on S. The order of lifting can affect the resulting constraint. Finding sequence-independent liftings is a topic of active research in integer programming. For more details see [42].

1.7   Semidefinite Constraints

Let $\xi = (x, y, w) \in \mathbb{R}^{n+2m}$. By taking products of the constraints defining (1.1), we can obtain nonconvex quadratic constraints on the elements of ξ. For example, the complementarity relationships imply $y_i w_i = 0$ for each i. These constraints can be relaxed to linear constraints by introducing a matrix $\varUpsilon$ to represent

$$\varUpsilon = \xi \xi^T,$$

replacing all quadratic terms by the corresponding entries in $\varUpsilon$ and then relaxing the equality to the semidefinite inequality

$$\varUpsilon \succeq \xi \xi^T.$$

This leads to a semidefinite programming (SDP) relaxation of (1.1) that is tighter than the LP relaxation. This approach is well-known for quadratically constrained quadratic programs, and there has been recent research on trying to improve it; see Luo et al. [33] and its references, for example.

Tightened linear equalities for (1.1) can be obtained by projecting down from the $(\varUpsilon, \xi)$ space onto the ξ space. This convex relaxation procedure has been extensively analyzed for 0–1 programming, and it has been shown by Lovasz and Schrijver [32] that repeated application leads to the convex hull of the feasible region. Kojima and Tunçel [30] explore SDP relaxations of quadratic constraints in detail and characterize the results of successive convex relaxation. They extend the work of [32] to finding the convex hull of a general nonconvex region, not necessarily one that arises from a 0 to 1 integer programming problem. Variants of their procedure converge to the convex hull in a finite number of steps. Anstreicher [3] showed that an SDP approach can be complementary to using the reformulation–linearization technique. Recent work on semidefinite relaxation approaches to mixed integer nonlinear programming problems includes [19,41].

We experimented with using CSDP [17] for solving SDP relaxations of LPCCs generated in the same way as those in Sect. 1.2. We were unable to solve instances as large as those in Sect. 1.2 in reasonable computational times, so we report results on smaller instances. We used a positive semidefinite matrix of the form

$$\bar{\varUpsilon} \triangleq \begin{bmatrix} 1 \\ x \\ y \end{bmatrix} \begin{bmatrix} 1 & x^T & y^T \end{bmatrix} = \begin{bmatrix} 1 & x^T & y^T \\ x & xx^T & xy^T \\ y & yx^T & yy^T \end{bmatrix},$$

Table 1.6 SDP results

			Percentage of gap closed	
n	m	k	SOCP	SDP
20	50	50	15.2	75.0
30	40	60	24.3	63.4
40	30	60	26.7	69.3
40	40	10	22.6	54.0
50	30	20	54.2	99.9
60	20	30	76.9	100.0

with the equality constraint relaxed. The model included constraints that all entries in Υ be nonnegative, that the entries corresponding to $y_i w_i$ be zero, that the entries in the first row and column of $\bar{\Upsilon}$ satisfy the appropriate linear constraints on x and y in (1.1), that the linear combinations of entries in $\bar{\Upsilon}$ corresponding to the products $y_i(\bar{A}x + \bar{B}y + \bar{C}w - b)_j$, $(\bar{A}x + \bar{B}y + \bar{C}w - b)_i(\bar{A}x + \bar{B}y + \bar{C}w - b)_j$ and $x_i(\bar{A}x + \bar{B}y + \bar{C}w - b)_j$ be nonnegative, and that the entries in $\bar{\Upsilon}$ corresponding to the terms x_i, x_i^2, y_j, and y_j^2 satisfy the convex quadratic constraints

$$x_i^2 \leq x_i^{\mathrm{UB}} x_i \quad \text{and} \quad y_j^2 \leq y_j^{\mathrm{UB}} y_j.$$

The results are contained in Table 1.6. Also contained in the table are results for the second-order cone programming (SOCP) relaxation which includes all the constraints of the SDP, except for the requirement that Υ be positive semidefinite. A lower bound was obtained by solving the relaxations of these problems containing the bound cuts (1.2) and the McCormick constraints (1.18), and the integer program (1.6) was solved in order to obtain the optimal value. The table gives the proportion of the gap between the optimal value and the lower bound from the McCormick cuts that is closed using the SOCP and SDP relaxations.

It is clear from the table that the SDP relaxation can be very strong. However, the computational time for this approach is not competitive with an LP-based integer programming approach, at least for these problems and when solving the SDPs to optimality with a primal–dual method. It may be helpful to use alternative techniques to solve the SDP problems, such as those in, for example, [18,24,31,47,48], techniques that can also be used to solve the SDPs approximately. Approximate solutions may be appropriate when the solver is incorporated into a branching scheme. It may also be more effective to add the linear constraints on Υ selectively as cutting planes.

It is also possible to construct constraints that the linear combinations of entries in $\bar{\Upsilon}$ corresponding to the products $y_i w_k$, $w_i w_k$, $w_i(\bar{A}x + \bar{B}y + \bar{C}w - b)_j$, and $x_i w_k$ be nonnegative, but we found these constraints resulted in SDPs that were too large for our solver. In principle, these constraints could be added as cutting planes, as could the earlier ones.

1.8 Conclusions

The cuts described in this chapter can dramatically reduce the gap between the global optimal value and the lower bound provided by a simple linear programming relaxation. We investigated adding whole families of cuts and quantifying how the lower bound is improved. The cuts can often be greatly improved by refining them, that is, by applying them and then recalculating them. For example, with bound cuts, a lower bound can be calculated using a single application of the bound-calculating procedure. The gap between the optimal value and this lower bound is reduced by 47% or more in all our test cases when the bounds are refined eight times, and the overall reduction in the duality gap is on the order of 75–90% when compared with the initial LP relaxation. Further, using refinements can often result in a lower overall runtime to solve the LPCC to optimality, even when taking the time for the refinements into account. The bound cuts appear to be more effective than the disjunctive cuts, which are expensive to calculate in their full form. Methods to speed up the calculation of disjunctive cuts certainly save time, but they appear to give cuts that are noticeably weaker than using the full disjunctive cut generation LP. When the dimension of the x variables is not too large, we have shown how to construct linear relaxations of the nonconvex quadratic constraint $y^T w \leq 0$ by expressing the constraint in terms of just the x and y variables. These McCormick constraints can be very effective, especially if they are refined. The refined cuts can typically reduce the gap by about 90% on our test problems. Novel quadratic constraints can be used to improve the McCormick cuts, but the improvement is not great for our test instances. Tighter relaxations can be obtained by using semidefinite relaxations, but these are currently expensive computationally to solve to optimality. Methods to approximately solve the SDP relaxations could be useful, as could methods using the SDP as a cut-generation mechanism in an LP approach as in [31, 41, 48].

We are currently investigating the use of these classes of constraints in a branch-and-cut algorithm for finding the global optimum to the LPCC. In such an algorithm, the cuts are added more selectively, rather than adding whole families of cuts.

Acknowledgements We are grateful to two anonymous referees for their careful reading of the manuscript and constructive comments. The work of the first author was supported by the National Science Foundation under grant DMS-0715446 and by the Air Force Office of Sponsored Research under grants FA9550-08-1-0081 and FA9550-11-1-0260. The work of the second author was supported by the National Science Foundation grant CMMI-0969600 and by the Air Force Office of Sponsored Research under grants FA9550-08-1-0061 and FA9550-11-1-0151.

References

1. Adjiman, C.S., Androulakis, I.P., Floudas, C.A.: A global optimization method, αBB, for general twice-differentiable constrained NLPs—I. Theoretical advances. Comput. Chem. Eng. **22**(9), 1137–1158 (1998)

2. Al-Khayyal, F.A., Falk, J.E.: Jointly constrained biconvex programming. Math. Oper. Res. **8**(2), 273–286 (1983)
3. Anstreicher, K.M.: Semidefinite programming versus the reformulation-linearization technique for nonconvex quadratically constrained quadratic programming. J. Global Optim. **43**(2–3), 471–484 (2009)
4. Audet, C., Haddad, J., Savard, G.: Disjunctive cuts for continuous linear bilevel programming. Optim. Lett. **1**(3), 259–267 (2007)
5. Audet, C., Savard, G., Zghal, W.: New branch-and-cut algorithm for bilevel linear programming. J. Optim. Theory Appl. **38**(2), 353–370 (2007)
6. Balas, E.: Intersection cuts—a new type of cutting planes for integer programming. Oper. Res. **19**, 19–39 (1971)
7. Balas, E.: Disjunctive programming. Ann. Disc. Math. **5**, 3–51 (1979)
8. Balas, E.: Disjunctive programming: properties of the convex hull of feasible points. Disc. Appl. Math. **89**(1–3), 3–44 (1998)
9. Balas, E., Ceria, S., Cornuéjols, G.: Mixed 0–1 programming by lift-and-project in a branch-and-cut framework. Manage. Sci. **42**(9), 1229–1246 (1996)
10. Balas, E., Ceria, S., Cornuéjols, G., Natraj, N.: Gomory cuts revisited. Oper. Res. Lett. **19**, 1–9 (1996)
11. Balas, E., Perregaard, M.: Lift-and-project for mixed 0-1 programming: recent progress. Disc. Appl. Math. **123**(1–3), 129–154 (2002)
12. Balas, E., Perregaard, M.: A precise correspondence between lift-and-project cuts, simple disjunctive cuts, and mixed integer Gomory cuts for 0-1 programming. Math. Program. **94**(2–3), 221–245 (2003)
13. Bao, X., Sahinidis, N.V., Tawarmalani, M.: Multiterm polyhedral relaxations for nonconvex, quadratically-constrained quadratic programs. Optim. Methods Softw. **24**(4–5), 485–504 (2009)
14. Belotti, P.: Disjunctive cuts for nonconvex MINLPs. Technical report, Lehigh University, Bethlehem (2009)
15. Belotti, P., Lee, J., Liberti, L., Margot, F., Wachter, A.: Branching and bounds tightening techniques for non-convex MINLP. Optim. Methods Softw. **24**(4–5), 597–634 (2009)
16. Belotti, P., Liberti, L., Lodi, A., Nannicini, G., Tramontani, A.: Disjunctive inequalities: applications and extensions. In: Cochran, J.J. (ed.) Encyclopedia of Operations Research and Management Science. Wiley (2010)
17. Borchers, B.: CSDP, a C library for semidefinite programming. Optim. Methods Softw. **11**, 613–623 (1999)
18. Burer, S., Monteiro, R.D.C.: A nonlinear programming algorithm for solving semidefinite programs via low-rank factorization. Math. Program. **95**(2), 329–357 (2003)
19. Burer, S., Saxena, A.: Old wine in a new bottle: the MILP road to MIQCP. Technical report, Department of Management Sciences, University of Iowa, Iowa City (2009)
20. Byrd, R., Nocedal, J., Waltz, R.: KNITRO: an integrated package for nonlinear optimization. In: di Pillo, G., Roma, M. (eds.) Large-Scale Nonlinear Optimization, pp. 35–59. Springer, Berlin (2006)
21. Ceria, S., Pataki, G.: Solving integer and disjunctive programs by lift-and-project. In: Proceedings of the Sixth IPCO Conference. Lecture Notes in Computer Science, vol. 1412, pp. 271–283. Springer, Berlin (1998)
22. de Farias Jr., I.R., Johnson, E.L., Nemhauser, G.L.: Facets of the complementarity knapsack polytope. Math. Oper. Res. **27**(1), 210–226 (2002)
23. Fletcher, R., Leyffer, S.: Solving mathematical programs with complementarity constraints as nonlinear programs. Optim. Methods Softw. **18**(1), 15–40 (2004)
24. Helmberg, C.: Numerical evaluation of SB method. Math. Program. **95**(2), 381–406 (2003)
25. Hobbs, B.F., Metzler, C.B., Pang, J.S.: Strategic gaming analysis for electric power systems: an MPEC approach. IEEE Trans. Power Syst. **15**(2), 638–645 (2000)
26. Hu, J., Mitchell, J.E., Pang, J.S.: An LPCC approach to nonconvex quadratic programs. Math. Program. **133**(1–2), 243–277 (2012)

27. Hu, J., Mitchell, J.E., Pang, J.S., Bennett, K.P., Kunapuli, G.: On the global solution of linear programs with linear complementarity constraints. SIAM J. Optim. **19**(1), 445–471 (2008)
28. Hu, J., Mitchell, J.E., Pang, J.S., Yu, B.: On linear programs with linear complementarity constraints. Technical report, Department of Mathematical Sciences, Rensselaer Polytechnic Institute, Troy (2009). Accepted for publication in J. Global Optim.
29. Júdice, J.J., Sherali, H.D., Ribeiro, I.M., Faustino, A.M.: A complementarity-based partitioning and disjunctive cut algorithm for mathematical programming problems with equilibrium constraints. J. Global Optim. **36**, 89–114 (2006)
30. Kojima, M., Tuncel, L.: Cones of matrices and successive convex relaxations of nonconvex sets. SIAM J. Optim. **10**(3), 750–778 (2000)
31. Krishnan, K., Mitchell, J.E.: A unifying framework for several cutting plane methods for semidefinite programming. Optim. Methods Softw. **21**(1), 57–74 (2006)
32. Lovász, L., Schrijver, A.: Cones of matrices and set-functions and 0-1 optimization. SIAM J. Optim. **1**(2), 166–190 (1991)
33. Luo, Z.-Q., Ma, W.-K., So, A.M.-C., Ye, Y., Zhang, S.: Semidefinite relaxation of quadratic optimization problems. IEEE Signal Process. Mag. **27**(3), 20–34 (2010)
34. McCormick, G.P.: Computability of global solutions to factorable nonconvex programs: part I—convex underestimating problems. Math. Program. **10**, 147–175 (1976)
35. Mitchell, J.E., Pang, J.S., Yu, B.: Convex quadratic relaxations of nonconvex quadratically constrained quadratic programs. Technical report, Mathematical Sciences, Rensselaer Polytechnic Institute, Troy (2011)
36. Nemhauser, G.L., Wolsey, L.A.: Integer and Combinatorial Optimization. Wiley, New York (1988)
37. Pang, J.S.: Three modeling paradigms in mathematical programming. Math. Program. **125**(2), 297–323 (2010)
38. Pang, J.S., Olson, T., Priebe, C.: A likelihodd-MPEC approach to target classification. Math. Program. **96**(1), 1–31 (2003)
39. Pang, J.S., Su, C.L.: On estimating pure characteristics models. Technical report, Industrial and Enterprise Systems Engineering, University of Illinois, Urbana-Champaign (in preparation)
40. Perregaard, M.: Generating disjunctive cuts for mixed integer programs. PhD Thesis, Carnegie Mellon University, Graduate School of Industrial Administration, Pittsburgh (2003)
41. Qualizza, A., Belotti, P., Margot, F.: Linear programming relaxations of quadratically constrained quadratic programs. In: IMA Volume Series. Springer (2010). Accepted. Tepper Working Paper 2009-E20 (revised 4/2010)
42. Richard, J.-P.P., Tawarmalani, M.: Lifted inequalities: a framework for generating strong cuts for nonlinear progams. Math. Program. **121**(1), 61–104 (2010)
43. Sahinidis, N.: BARON: a general purpose global optimization software package. J. Global Optim. **8**, 201–205 (1996)
44. Saxena, A.: Integer programming, a technology. PhD Thesis, Tepper School of Business, Carnegie Mellon University (2007)
45. Saxena, A., Bonami, P., Lee, J.: Convex relaxations of non-convex mixed integer quadratically constrained programs: extended formulations. Math. Program. **124**(1–2), 383–411 (2010)
46. Saxena, A., Bonami, P., Lee, J.: Convex relaxations of non-convex mixed integer quadratically constrained programs: projected formulations. Math. Program. **126**(2), 281–314 (2011)
47. Sivaramakrishnan, K.K.: A parallel interior point decomposition algorithm for block angular semidefinite programs. Comput. Optim. Appl. **46**(1), 1–29 (2010)
48. Sivaramakrishnan, K.K., Mitchell, J.E.: Properties of a cutting plane method for semidefinite programming. Technical report, Mathematical Sciences, Rensselaer Polytechnic Institute, Troy (2011)
49. Stubbs, R.A., Mehrotra, S.: A branch-and-cut method for 0-1 mixed convex programming. Math. Program. **86**, 515–532 (1999)
50. Tawarmalani, M., Sahinidis, N.: Convexification and Global Optimization in Continuous and Mixed-Integer Nonlinear Programming: Theory, Algorithms, Software, and Applications. Kluwer, Dordrecht (2002)

Chapter 2
Distributed Optimization in Networking: Recent Advances in Combinatorial and Robust Formulations

Minghua Chen and Mung Chiang

2.1 Introduction

Optimization has become an essential modeling language and design method for communication networks. It has been widely applied to many key problems, including power control, coding, scheduling, routing, congestion control, content distribution, and pricing. It has also provided a fresh angle to view the interactions across a network protocol stack as the solutions to an underlying optimization problem. A unique requirement for optimization in networks is that the solution algorithm must be distributed. This has in turn motivated the development of new tools in distributed optimization. Many of these results have been well documented. In this chapter, we turn to a sample of three recent results on some of the challenging new issues, centered around the need to tackle combinatorial or robust optimization formulation through distributed algorithms.

2.1.1 Garg–Konemann Framework for Fractional Packing Linear Programming Problems

A number of system design problems can be formulated as fraction packing linear programming problems. Such problems often come with exponential number of variables, and are challenging to solve. The example we will focus on in Sect. 2.2

M. Chen (✉)

Department of Information Engineering, The Chinese University of Hong Kong, Hong Kong
e-mail: minghua@ie.cuhk.edu.hk

M. Chinag
Department of Electrical Engineering, Princeton University, Princeton, NJ, USA
e-mail: chiangm@princeton.edu

T. Terlaky and F.E. Curtis (eds.), *Modeling and Optimization: Theory and Applications*, Springer Proceedings in Mathematics & Statistics 21, DOI 10.1007/978-1-4614-3924-0_2, © Springer Science+Business Media, LLC 2012

is peer-to-peer (P2P) streaming capacity problem [1–3]: given a P2P network with a source node and a set of receivers, how to embed a set of multicast trees spanning the receivers and to determine the amount of flow in each tree, such that the sum of flows over these trees is maximized?

Garg and Konemann [4] presented a framework for solving these challenging problems approximately in polynomial time. The intuitive observation behind the framework is similar to those of column generation [5–7]: although solving the problem exactly may require exploring the space spanned by all the exponential number of variables, solving the problem approximately only requires tuning a polynomial-size subset of them. The column generation technique does not specify how to find one such subset of the variables, as they are usually problem-specific. Interestingly, for all fractional packing problems, the Garg–Konemann framework provides a systematic way to find one such subset of the variables.

2.1.2 Markov Approximation for Combinatorial Optimization

Many important network resource allocation problems are combinatorial in nature. The objective of these problems is to maximize a system-wide performance metric, which involves enumerating all possible configurations of many independent entities. Problems of this kind appear in various domains, including wireless networking [8–10], P2P networking [2, 3], content distribution [11], and cloud computing [12, 13].

We observe a gap between the known and the desired solutions for many of these combinatorial problems. On one hand, exact solutions are computationally prohibitive. Research thus focuses on finding efficient approximation algorithms. However, most of these algorithms only allow *centralized* implementations.

On the other hand, system designers usually prefer distributed algorithms. Distributed algorithms are more adaptable to network resource fluctuation and users joining and leaving as compared to centralized ones.

In Sect. 2.3, we present a general Markov approximation technique that allows us to approximately solve combinatorial optimization problems in a distributed manner. This also addresses the computational complexity issue to a certain extent because the distributed implementation often allows parallel processing by different elements. To demonstrate the usefulness of the technique, we apply it to design a distributed streaming algorithm for P2P systems that achieves close-to-optimal performance.

2.1.3 Distributed Robust Optimization

Robustness of optimization models for network problems in communication networks has been an under-explored topic. Most existing algorithms for solving robust optimization problems are centralized, thus not suitable for networking problems that demand distributed solutions. In Sect. 2.4, we summarize a first step toward a systematic theory for designing distributed *and* robust optimization models and algorithms. We first discuss several models for describing parameter uncertainty sets that can lead to decomposable problem structures and thus distributed solutions. These models include ellipsoid, polyhedron, and D-norm uncertainty sets. We then apply these models in solving a robust rate control problem in wireline networks.

2.2   P2P Streaming Capacity

2.2.1   *Problem Settings*

Consider a P2P network as a P2P graph $G = (V, E)$, where each node $v \in V$ may be the source, a receiver, or a helper that serves only as a relay; each edge $e = (u, v) \in E$ represents a neighboring relationship between vertices u and v. We assume that data rate bottlenecks only appear at node uplinks. Denoting by $C(v)$ the uplink capacity of node v, we have for each node v,

$$\sum_{u \in V} x_{vu} \leq C(v), \qquad (2.1)$$

where x_{vu} is the rate node v transmits to node u. This assumption is widely adopted in the P2P literature because in today's Internet, access links are the bottlenecks rather than backbone links, and uplink capacity is several times smaller than downlink capacity in access networks.

We consider the single session scenario in this chapter.[1] In this scenario, the session content originates from one source s and is distributed to a given set of receivers R, possibly using a set of helpers H. A packet in the session stream starts from the source s, and traverses over all nodes in R, and some nodes in H—the traversed paths form a Steiner tree in the overlay graph G. Here a Steiner tree refers to a tree that connects the sender and all the receivers and allows the sender to reach any receiver over it. Different packets in the same session may traverse different trees, and we call each tree a *sub-tree*, and call their superposition a *multi-tree*.

There are P2P protocol constraints on sub-trees. The most frequently encountered one is degree constraint. For example, in the widely used P2P protocol of BitTorrent, although one node has 30–50 neighbors in G, it can upload to at most five of them as peers. This gives an outgoing degree bound for each node and constrains the construction of the trees. Degree bounds always apply to receivers and helpers and,

---

[1]As compared to single-session scenario, the multi-session scenario involves multiple sessions competing for the underlying physical link capacities, and one has to take the fairness consideration into account when formulating the problem. We refer interested readers on the multi-session study to [1].

in some scenarios, to the source as well. Let $m_{v,t}$ be the number of outgoing edges of node v in tree t, and the bound be $M(v)$: $m_{v,t} \leq M(v), \forall t$. We denote by T the set of all allowed sub-trees: trees that satisfy the constraints such as the degree bounds.

For each tree $t \in T$, we denote by y_t the rate of the sub-stream supported by this tree. We say the session has a rate $r = \sum_{t \in T} y_t$ if all the receivers obtain the streaming contents at a rate of r or above. A rate is called *achievable* if there is a multi-tree in which all trees satisfy the topology constraint ($t \in T$) and transmission rates satisfy the uplink capacity constraint. We define *P2P streaming capacity* as the largest achievable rate.

2.2.2 Formulation

The P2P streaming capacity problem for single session is formulated as follows.

2.2.2.1 Single-Session (Primal) Problem

$$\text{maximize} \quad r = \sum_{t \in T} y_t \tag{2.2}$$

$$\text{subject to} \quad \sum_{t \in T} m_{v,t} y_t \leq C(v), \forall v \in V \tag{2.3}$$

$$y_t \geq 0, \forall t \in T \tag{2.4}$$

$$\text{variables} \quad y_t, \forall t \in T \tag{2.5}$$

For those trees not selected in the solution, their rates y_t are simply zero.

The dual problem associates a nonnegative variable $p(v)$, interpreted as price, with each node v corresponding to constraint (2.3). It can be derived to be the following problem.

2.2.2.2 Single-Session Dual Problem

$$\text{minimize} \quad \sum_{v \in V} C(v) p(v) \tag{2.6}$$

$$\text{subject to} \quad \sum_{v \in V} m_{v,t} p(v) \geq 1, \forall t \in T, \tag{2.7}$$

$$p(v) \geq 0, \forall v \in V \tag{2.8}$$

$$\text{variables} \quad p(v), \forall v \in V \tag{2.9}$$

We can interpret the dual problem this way: $p(v)$ is the per unit flow price for any edge outgoing from v. If node v uploads with full capacity, the incurred cost is $p(v)C(v)$. There are $m_{v,t}$ connections outgoing from node v in tree t, and thus the *total tree price* for tree t, which is defined as the sum of prices in any edge in tree t, is $\sum_{v \in V} m_{v,t} p(v)$. Therefore, the dual problem is to minimize the total full capacity tree cost given that the tree price is at least 1, and the minimization is over all possible **p**, where $\mathbf{p} := \{p(v), \forall v \in V\}$ is the price vector. For notational simplicity, we use $p(\cdot) : V \to R^+ \cup \{0\}$ to represent **p**.

In general, the number of trees we need to search when computing the right multi-tree grows exponentially in the size of the network. This dimensionality increase is the consequence of turning a difficult graph-theoretic, discrete problem into a continuous optimization problem. Hence, the primal problem can have possibly exponential number of variables and its dual can have an exponential number of constraints, neither of which is suitable for direct solution in polynomial time. However, as detailed in the next subsection, the above representations turn out to be very useful to allow a primal–dual update outer loop that converts the combinatorial problem of multi-tree construction into a usually simpler problem of smallest price tree (SPT) construction.

2.2.3 Algorithm and Performance

Adapting the technique for solving the maximum multi-commodity flow problem in [4], we design an iterative combinatorial algorithm that solves the primal and dual problems approximately, where tree-flows are augmented in the primal solution and dual variables are updated iteratively. Our algorithm constructs peering multi-trees that achieve an objective function value within $(1 + \zeta)$-factor of optimal.

For a given tree t and prices $p(\cdot)$, let $Q(t, p)$ denote the left-hand side (LHS) of constraint (2.7), which we call the *price* of tree t. A set of prices $p(\cdot)$ is a feasible solution for the dual program if and only if

$$\min_{t \in T} Q(t, p) \geq 1.$$

The algorithm works as follows. Start with initial weights $p(v) = \frac{\delta}{C(v)}$ for all $v \in V$. Parameter δ depends on ζ and is described in more detail later. Repeat the following steps until the dual objective function value becomes greater than 1:

1. Compute a tree $\bar{t}$ for which $Q(t, p)$ is minimum. We call $\bar{t}$ a *SPT* problem, algorithms for which are developed in next section.
2. Send the maximum flow on this tree $\bar{t}$ such that uplink capacity of at least one internal node is saturated. Let $I(t)$ be the set of internal nodes in tree t. The flow sent on this tree is

$$y = \min_{v \in I(\bar{t})} \frac{C(v)}{m_{v,\bar{t}}}. \tag{2.10}$$

Fig. 2.1 The primal–dual
algorithm for single-session
P2P streaming capacity
computation

Primal-Dual Algorithm: Single-Session

$p(v) \leftarrow \frac{\delta}{C(v)}, flow(v) \leftarrow 0, \forall v \in V, \; Y \leftarrow 0,$
$D \leftarrow 0$

while $D < 1$
    Pick tree $t \in T$ with the smallest $Q(t,p)$
    $y \leftarrow \min_{v \in I(t)} C(v)/m_{v,t}$
    $\bar{t} \leftarrow \arg \min_{v \in I(t)} C(v)/m_{v,t}$
    $flow(v) \leftarrow flow(v) + y m_{v,\bar{t}}, \forall v \in I(\bar{t})$
    $Y \leftarrow Y + y$
    $p(v) \leftarrow p(v)(1 + \epsilon \frac{m_{v,\bar{t}} y}{C(v)})$
    $D \leftarrow \sum_{v \in V} C(v) p(v)$
end while

Compute scaling factor $\alpha \leftarrow \max_{v \in V} \frac{flow(v)}{C(v)}$;
Output capacity $r^* \leftarrow Y/\alpha$;

---

3. Update the prices $p(v)$ as

$$p(v) \leftarrow p(v) \left(1 + \frac{\varepsilon m_{v,\bar{t}} y}{C(v)}\right), \forall v \in I(\bar{t}).$$

    where ε depends on θ and is explained in more detail later.

4. Increment the flow Y sent so far by y.

The optimality gap can be estimated by computing the ratio of the primal and dual objective function values in each step of the above iteration, which can be terminated after the desired proximity to optimality is achieved. When the above iteration terminates, primal capacity constraints on each uplink may be violated, since we were working with the original (and not residual) uplink capacities at each stage. To remedy this, we scale down the flows uniformly so that uplink capacity constraints are satisfied.

The pseudo-code for the above procedure is provided in Fig. 2.1. Array $flow(v)$ keeps track of the traffic on uplink of node v as the algorithm progresses. The dual objective function value is tracked by variable D which is initialized to 0. After the "while" loop terminates, the maximum factor by which the uplink capacity constraint is violated on any uplink is computed as α, which divides the total flow Y, and the resulting value is output as r^*.

The theorem below states accuracy and complexity properties of the algorithm. Its proof can be found in [1].

Theorem 2.1. *For any given $\zeta > 0$, the Single-Session Primal–Dual Algorithm computes a solution with objective function value within $(1 + \zeta)$-factor of the optimum, for algorithmic parameters $\varepsilon(\zeta) = 1 - \frac{1}{\sqrt{1+\zeta}}$ and $\delta(\zeta) = \frac{1+\varepsilon}{[(1+\varepsilon)|V|]^{1/\varepsilon}}$.*

It runs in time polynomial in the input size and $\frac{1}{\varepsilon}$: $O\left(\frac{|V|\log|V|}{\varepsilon^2}T_{spt}\right)$, where T_{spt} is the time to compute a SPT.

This unifying primal–dual framework works for *all* of the single session problems. The core issue now lies with the inner loop of SPT computation: can this be accomplished in polynomial time for a given price vector? This graph-theoretic problem is generally much more tractable than the original problem of searching for the multi-tree that maximizes the achievable rate. However, when the given graph G is not a full mesh, or when there are degree bounds on nodes in each tree, or when there are helper nodes, computing an SPT becomes difficult.

2.2.4 Computing SPT

For iterative algorithms, efficiently and accurately computing an SPT for a given set of node prices is the key module in each outer loop. We have developed SPT algorithms for general problem formulations over P2P graph in [1]. We present the case of general P2P graph without degree bound and with helpers in the following; we leverage special structures of the P2P graph G in the development.

We now study the cases where the given graph G is not complete and without degree bounds in the trees. We consider the case with the presence of helpers. In this case, the SPT computation problem is a minimum cost directed Steiner tree problem *with symmetric connectivity and a special structure on the costs*—the costs of all edges going out of a node are equal. We leverage these two special features to accelerate the algorithm through a graph transformation.

Let $[v_1,v_2]$ represent the pair of links connecting two neighboring nodes v_1 and v_2. We want to find the minimum price directed Steiner tree connecting source s and a set of receivers in R. We transform the directed graph G to an undirected graph $G' = (V',E')$, which represents the adjacency between link pairs and the source node s:

- For source node s, we copy it into G'.
- For every two neighboring nodes $v_1,v_2 \in V$, we map the link pair $[v_1,v_2]$ to a node $n_{[v_1,v_2]}$ in G'.
- For every node $v_1 \in V$ in G, we map it to a series of *undirected* links connecting nodes $n_{[v_1,v_2]}$ and $n_{[v_1,v_3]}$ in G' where v_2 and v_3 are any neighbors of v_1 in G; we set the prices of these undirected links to $p(v)$.
- In graph G', we connect any two of nodes s and $n_{[s,v]}$, with a series of *undirected* links, where v is any neighbor of s in G; we set the prices of these links to be $p(s)$.

An example of such transformation is illustrated in Fig. 2.2.

For the undirected graph G', we consider the following group Steiner tree problem. For every receiver $r \in R$ in G, we group all the nodes $n_{[r,v]} \in V'$, $v \in V$, into a set, denoted by g_r. Set g_r in G' corresponds to the set of link pairs in G connecting

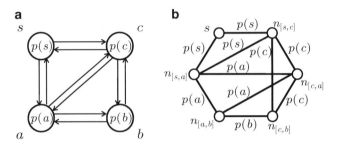

Fig. 2.2 (**a**) An overlay graph with source node s, receiver nodes a and b, and Steiner node c with node prices being p_s, p_a, p_b, and p_c respectively; (**b**) the undirected graph mapped from the overlay graph in (**a**); for example, the link pair between a and b in (**a**) maps to the node $n_{[a,b]}$ in (**b**), and node a maps to three links with link cost p_a in (**b**)

r to all its neighbors. We also construct a set g_s that contains only the source s in G'. The group Steiner tree problem in G' is to find the minimum price Steiner tree that connects at least one node from each of sets g_s and g_r, $r \in R$. Solving the group Steiner tree problem in undirected graph is NP-hard [14]. The authors in [14] propose a polynomial time algorithm that achieves an approximation factor of $\frac{1}{O(\ln N_g \ln^3 N_m)}$, where N_g is the number of groups and N_m is total number of nodes.

The following theorem, proved in [1], states that finding the minimum cost Steiner tree in G is equivalent to searching the minimum cost group Steiner tree in G'.

Theorem 2.2. *Consider finding a Steiner tree in G that connects s and all nodes in R, and searching a group Steiner tree in G' that connects at least one node from each of node sets g_s and g_r, $r \in R$, the following is true:*

1. *A directed Steiner tree in G can be mapped to a group Steiner tree in G' with the same price in polynomial time.*
2. *A group Steiner tree in G' can be mapped to a directed Steiner tree in G with the same or less price in polynomial time.*

Consequently, the optimal group Steiner tree in G' can be mapped to the optimal directed Steiner tree in G in polynomial time and vice versa. Furthermore, their prices are equal.

The minimum cost directed Steiner tree problem is hard to approximate to a factor better than $\frac{1}{\ln |R|}$ [15]. An $\frac{1}{O(|R|^\varepsilon)}$-factor approximation algorithm that runs in polynomial time for any fixed $\varepsilon > 0$ is given in [16]. Theorem 2.2 states that the directed Steiner tree problem in graph G can be approached by studying a group Steiner tree problem in G'. We first apply the randomized algorithm proposed in [14] to G', and get a group Steiner tree with an approximation factor of $\frac{1}{O(\ln |R| \ln^3 N_m)}$. N_m corresponds to total number of link pairs in G, and is at most $|V|^2$. We then map this group Steiner tree to a directed Steiner tree in G. Since this mapping keeps or reduces the price, at the end we compute a directed Steiner tree in G with an approximation factor of $\frac{1}{O(\ln|R|\ln^3|V|)}$ in polynomial time. In the case where $|V| = O(|R|)$, we can compute a directed Steiner tree with an approximation factor of $\frac{1}{O(\ln^4|R|)}$ in polynomial time.

2.3 Markov Approximation for Combinatorial Optimization

2.3.1 General Framework

2.3.1.1 Problem Formulation

Consider a system with a set of users R, and a set of configurations $\mathcal{F}$. A network configuration $f \in \mathcal{F}$ consists of individual components using one of its local configurations. The system obtains a certain performance when it operates under configuration f, denoted by W_f. The problem of maximizing the system performance by choosing the best configuration can then be cast as the following combinatorial optimization problem

$$\mathbf{MWC} : \max_{f \in \mathcal{F}} W_f. \tag{2.11}$$

One well-known example is the neighbor selection problem in P2P streaming systems. The name of the game is to assign the "best" set of neighbors for every peer/cache, in order to form the best topology to support a streaming video at the highest possible quality [2, 3]. In the example, a local configuration of a peer is the set of neighbors it connects to, and a network configuration is a vector consisting of local configurations of all the peers. The system performance metric in the example is the video streaming rate.

For the problem **MWC**, an equivalent formulation is

$$\mathbf{MWC} - \mathbf{EQ} : \max_{p \geq 0} \sum_{f \in \mathcal{F}} p_f W_f \tag{2.12}$$

$$\text{s.t.} \ \sum_{f \in \mathcal{F}} p_f = 1,$$

where p_f is the percentage of time the configuration f is in use. Treating W_f in (2.11) as the "weight" of f, the problem **MWC** is to find a maximum weighted configuration.

2.3.1.2 Log–Sum–Exp Approximation

To gain insights on the structure of the problem **MWC**, we approximate the max function in (2.11) by a differentiable function as follows:

$$\max_{f \in \mathcal{F}} W_f \approx \frac{1}{\beta} \log \left(\sum_{f \in \mathcal{F}} \exp \left(\beta W_f \right) \right) \triangleq g_\beta \left(W \right), \qquad (2.13)$$

where β is a positive constant and $W \triangleq \left[W_f, f \in \mathcal{F} \right]$. This approximation is known as the convex log–sum–exp approximation to the max function. The following bound on approximation accuracy is well known.

Theorem 2.3. *For a positive constant β and n nonnegative real variables $y_1, y_2, \ldots, y_n$, we have*

$$\max(y_1, \ldots, y_n) \leq \frac{1}{\beta} \log \left(\exp(\beta y_1) + \cdots + \exp(\beta y_n) \right)$$

$$\leq \max(y_1, \ldots, y_n) + \frac{1}{\beta} \log n. \qquad (2.14)$$

Hence, $\max(y_1, \ldots, y_n) = \lim_{\beta \to \infty} \frac{1}{\beta} \log \left(\exp(\beta y_1) + \cdots + \exp(\beta y_n) \right).$

We summarize some important, well-known observations of $g_\beta(x)$ in the following theorem. Some of these observations were also found relevant in the context of Geometric Programming [17].

Theorem 2.4. *For the log–sum–exp function $g_\beta(W)$, we have*

- *Its conjugate function is given by*

$$g_\beta^* (p) = \begin{cases} \frac{1}{\beta} \sum_{f \in \mathcal{F}} p_f \log p_f & \text{if } p \geq 0 \text{ and } 1^T p = 1 \\ \infty & \text{otherwise.} \end{cases} \qquad (2.15)$$

- *It is a convex and closed function; hence, the conjugate of its conjugate $g_\beta^*(p)$ is itself, i.e., $g_\beta(W) = g_\beta^{**}(W)$. Specifically,*

$$g_\beta(W) = \max_{p \geq 0} \sum_{f \in \mathcal{F}} p_f W_f - \frac{1}{\beta} \sum_{f \in \mathcal{F}} p_f \log p_f \qquad (2.16)$$

$$\text{s.t.} \sum_{f \in \mathcal{F}} p_f = 1.$$

Several observations can be made. First, by the log–sum–exp approximation in (2.13), we are implicitly solving an approximated version of the problem **MWC – EQ**, off by an *entropy* term $-\frac{1}{\beta} \sum_{f \in \mathcal{F}} p_f \log p_f$. The optimality gap is thus bounded by $\frac{1}{\beta}\log|\mathcal{F}|$, where $|\mathcal{F}|$ represents the size of $\mathcal{F}$. This is a direct consequence of us theoretically approximating the max function by a log–sum–exp function in (2.13). Practically, adding this additional entropy term in fact opens new design space for exploration. Second, the approximation becomes exact as β approaches infinity. As discussed in [18], however, usually β should not take too large values as there are practical constraints or convergence rate concerns in the algorithm design afterwards. Third, we can derive a close form of the optimal solution of the problem in (2.16). Let λ be the Lagrange multiplier associated with the equality constraint in (2.16) and $p_f^*(x), f \in \mathcal{F}$ be the optimal solution of the problem in (2.16). By solving the Karush–Kuhn–Tucker (KKT) conditions [19] of the problem in (2.16):

$$W_f - \frac{1}{\beta}\log p_f^*(W) - \frac{1}{\beta} + \lambda = 0, \ \forall f \in \mathcal{F}, \tag{2.17}$$

$$\sum_{f \in \mathcal{F}} p_f^*(W) = 1, \ \lambda \ge 0, l \tag{2.18}$$

we have

$$p_f^*(W) = \frac{\exp\left(\beta W_f\right)}{\sum_{f' \in \mathcal{F}}\exp\left(\beta W_{f'}\right)}, \forall f \in \mathcal{F}. \tag{2.19}$$

By time-sharing among different configurations f according to $p_f^*(W)$, we can solve the problem **MWC − EQ**, and hence the problem **MWC**, approximately. We remark that the optimality gap is bounded by $\frac{1}{\beta}\log|\mathcal{F}|$, which can be made small by choosing large β.

2.3.1.3 Distributed Markov Chain Monte Carlo

We design a Markov chain with a state space being the set of all feasible peering configurations $\mathcal{F}$ and has a stationary distribution as $p_f^*(W)$ in (2.19). We implement the Markov chain to guide the system to optimize the configuration. As the system hops among configurations, the Markov chain converges and the configurations are time-shared according to the desired distribution $p_f^*(W)$.

The key lies in designing such Markov chain that allows distributed implementation. Since $p_f^*(W)$ in (2.19) is product-form, it suffices to focus on designing time-reversible Markov chains [18].

Let $f, f' \in \mathcal{F}$ be two states of Markov chain, and denote $q_{f,f'}$ as the transition rate from state f to f'. We have two degrees of freedom in designing a time-reversible Markov chain:

- *The state space structure*: we can add or cut direct transitions between any two states, given that the state space remains connected and any two states are reachable from each other. For example, assume that the four-states Markov

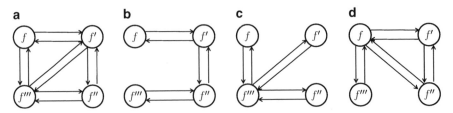

Fig. 2.3 The Markov chains in (**b**), (**c**), (**d**), by adding/removing transition edge-pair between two states in the time-reversible Markov chain in (**a**), are also time-reversible. All Markov chains have the same stationary distribution

chain in Fig. 2.3a is time-reversible. The "sparse" Markov chains in Fig. 2.3b–d, modified from the "dense" one in Fig. 2.3a by adding/removing transition edge-pair between two states, are also time-reversible. Furthermore, all Markov chains share the same stationary distribution.

- *The transition rates*: we can explore various options in designing $q_{f,f'}$, given that the detailed balance equation is satisfied, i.e.,

$$p_f^*(W)q_{f,f'} = p_f^*(W)q_{f',f}, \quad \forall f, f' \in \mathcal{F}. \tag{2.20}$$

Satisfying the above equations guarantees the designed Markov chain and has the desired stationary distribution as in (2.19). For instance, 802.11b wireless CSMA MAC protocol implements a time-reversible Markov chain with

$$q_{f,f'} = \alpha \exp\left(\beta\left(W_{f'} - W_f\right)\right), \tag{2.21}$$

$$q_{f',f} = \alpha, \tag{2.22}$$

where α is a positive constant [18]. We show another example on P2P streaming later in this chapter.

In Markov approximation framework, with large values of β, the system hops toward better configurations more greedily. However, this may as well lead to the system getting trapped in locally optimal configurations. Hence there is a trade-off to consider when setting the value of β. The value of β also affects the convergence rate of the time-reversible Markov chain to the desired stationary distribution. It is worth future investigation to further understand the impact of β.

2.3.2 Application to Distributed P2P Streaming

In this subsection, we apply Markov approximation technique elaborated above to design distributed algorithms for maximizing P2P streaming rate under node degree bounds.

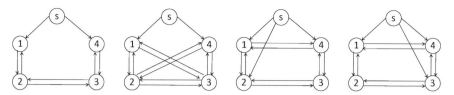

Fig. 2.4 Peering configuration examples for a five-node network with node degree bound 3 for each node

2.3.2.1 Settings and Notations

Like in Sect. 2.2, we model the P2P overlay network as a general directed graph $G = (V, E)$, where V denotes the set of nodes and E denotes the set of links. Each link in the graph corresponds to a TCP/UDP connection between two nodes. Each node $v \in V$ is associated with an upload capacity $C(v) \geq 0$. We assume there is no constraint on the downloading rate for each node $v \in V$. This assumption can be partly justified by the empirical observation that as residential broadband connections with asymmetric upload and download rates become increasingly dominant, bottlenecks typically are at the uplinks of the access networks rather than on end-user download speeds or in the middle of the Internet.

We again focus on the single-source streaming scenario, i.e., a source s broadcasts a continuous stream of contents to the entire network; we denote its receiver set as $R \triangleq V - \{s\}$.

We consider the peering constraints that each node has a degree bound B_v, i.e., it can only *simultaneously* connect to a B_v number of neighbors due to connection overhead cost. We allow different nodes to have different degree bounds. Figure 2.4 shows four sample peering configurations of a five-node network with node degree bound 3 for each node.

2.3.2.2 Problem Formulation and Our Approach

Let $\mathcal{F}$ denote the set of all feasible peering configurations over graph G under node degree bounds. For a configuration $f \in \mathcal{F}$, let x_f be the maximum achievable broadcast rate under f. The problem of maximizing broadcast rate under node degree bounds can be formulated as follows:

$$\mathbf{MRC} : \max_{f \in \mathcal{F}} x_f. \tag{2.23}$$

This problem is known to be NP-complete [2]. To address this problem in a distributed manner, we first develop a distributed broadcasting algorithm that can achieve x_f under arbitrary $f \in \mathcal{F}$. We then design a distributed topology-hopping algorithm that optimizes toward the best peering configurations. They operate in tandem to achieve a close-to-optimal broadcast rate under arbitrary node degree bounds, and over arbitrary overlay graph. We elaborate on the topology-hopping algorithm in the following. Details of the broadcasting algorithm are in [3].

Following Markov approximation framework, we need to design a Markov chain with a state space being the set of feasible configurations $\mathcal{F}$ and with a stationary distribution as follows:

$$p_f^*(x) = \frac{\exp(\beta x_f)}{\sum\limits_{f' \in \mathcal{F}} \exp(\beta x_{f'})}, \forall f \in \mathcal{F}. \tag{2.24}$$

Since $p_f^*(x)$ in (2.24) is product-form, it suffices to focus on designing time-reversible Markov chains. The key lies in designing such Markov chain that allows distributed implementation.

Let $f, f' \in \mathcal{F}$ be two states of Markov chain, and denote $q_{f,f'}$ as the transition rate from state f to f'. In our Markov chain design, we first specify its state space structure as follows: we set the transition rate $q_{f,f'}$ to be zero, unless f and f' satisfy that

- $|\mathcal{N}_f \cup \mathcal{N}_{f'} - \mathcal{N}_f \cap \mathcal{N}_{f'}| = 2$, i.e., f and f' differ by only two node pairs.
- There exists a node, denoted by v^*, so that $\mathcal{N}_f \cup \mathcal{N}_{f'} - \mathcal{N}_f \cap \mathcal{N}_{f'} \subseteq \{\{v^*, u\}, \forall u \in N_{v^*}\}$. That is, these two node pairs share a common node v^*.

In other words, we only allow direct transitions between two configurations if such transitions correspond to a single node swapping an in-use neighbor with a not-in-use one. Second, given the state space structure of Markov chain, we design the transition rates to favor distributed implementation while satisfying the detailed balance equation.

One possible option is to set $q_{f,f'}$ to be $\exp^{-1}(\beta x_f)$. One way to implement this option is for every node to generate a timer according to its *measured* receiving rate and counts down accordingly. When the timer expires, the dedicated node performs the neighbor swapping and resets its timer. As simple as the implementation may sound, this option is expensive to implement. Once the peering configuration changes, the system needs to notify all the nodes to measure the new receiving rate and reset their timers accordingly. It is not clear how to implement such system-wide notification in a low-overhead manner.

Instead, we choose to design $q_{f,f'}$ and $q_{f',f}$ as follows:

$$q_{f,f'} = \frac{1}{\exp(\tau)} \frac{\exp(\beta x_{f'})}{\exp(\beta x_{f'}) + \exp(\beta x_f)} \tag{2.25}$$

and

$$q_{f',f} = \frac{1}{\exp(\tau)} \frac{\exp(\beta x_f)}{\exp(\beta x_f) + \exp(\beta x_{f'})}, \tag{2.26}$$

where τ is a constant. It is straightforward to verify that detailed balance equation is satisfied. Such choices of transition rates do not require coordination or notification among peers in its implementation. We leave details of the distributed implementation to [3].

2.3.2.3 Impact of Inaccurate Observations

In practice, it is possible to obtain only an inaccurate measurement or estimate of x_f. These inaccuracies root in two sources. One is the noisy measurements of the maximum broadcast rates given the configuration. The other is the fast state transition of Markov chain, i.e., the Markov chain transits before the underlying broadcasting algorithm converges and thus it transits based on inaccurate observations of the broadcast rates.

Consequently, the topology hopping Markov chain may *not* converge to the desired stationary distribution $p_f^*(x)$. This observation motivates our following study on the convergence of Markov chain in the presence of inaccurate transition rates.

For each configuration $f \in \mathcal{F}$ with broadcast rate x_f, we assume its corresponding inaccurate observed rate belongs to the bounded region $\left[x_f - \Delta_f, x_f + \Delta_f \right]$, following arbitrary distribution. Δ_f is the inaccuracy bound and can be different for different f.

With such random inaccurate observed rates, it turns out that the topology-hopping process is still a time-reversible Markov chain [3]. We denote $\bar{p} : [\bar{p}_f(x), f \in \mathcal{F}]$ as its stationary distribution of the configurations in this Makrov chain. Recall that the stationary distribution of the configurations for the original inaccuracy-free topology hopping Markov chain is $p^* : [p_f^*(x), f \in \mathcal{F}]$. We use the total variance distance [20] to quantify the difference between p^* and $\bar{p}$, as

$$d_{TV}(p^*, \bar{p}) \triangleq \frac{1}{2} \sum_{f \in \mathcal{F}} |p_f^* - \bar{p}_f| \tag{2.27}$$

We have the following main result:

Theorem 2.5. *Let* $\Delta_{\max} = \max_{f \in \mathcal{F}} \Delta_f$, *and* $x_{\max} = \max_{f \in \mathcal{F}} x_f$. *The* $d_{TV}(p^*, \bar{p})$ *are bounded as follows:*

$$0 \leq d_{TV}(p^*, \bar{p}) \leq 1 - \exp(-2\beta\Delta_{\max}). \tag{2.28}$$

Further, the optimality gap in broadcast rates $|p^* x^T - \bar{p} x^T|$ *is bounded as below:*

$$|p^* x^T - \bar{p} x^T| \leq 2 x_{\max} (1 - \exp(-2\beta\Delta_{\max})). \tag{2.29}$$

The upper bound on $d_{TV}(p^*, \bar{p})$ shown in (2.28) is general, as it is independent of the number of configurations $|\mathcal{F}|$ and the distributions of inaccurate observed rates. The upper bound on $d_{TV}(p^*, \bar{p})$ shown in (2.28) decreases exponentially with the worst inaccuracy bound $\Delta_{\max}$ decreasing. It would be interesting to explore a tighter upper bound on $d_{TV}(p^*, \bar{p})$ than the one in (2.28).

2.4   Distributed Robust Optimization

2.4.1   General Framework

Despite the importance and success of using optimization theory to study communication and network problems, most work in this area makes the unrealistic
assumption that the data defining the constraints and objective function of the
optimization problem can be obtained precisely. We call the corresponding problems "nominal." In many practical problems, these data are typically inaccurate,
uncertain, or time-varying. Solving the nominal optimization problems may lead to
poor or even infeasible solutions when deployed.

Over the last 10 years, robust optimization has emerged as a framework of
tackling optimization problems under data uncertainty (e.g., [21–25]). The basic
idea of robust optimization is to seek a solution which remains feasible and near-
optimal under the perturbation of parameters in the optimization problem. Each
robust optimization problem is defined by three-tuple: *a nominal formulation, a
definition of robustness*, and *a representation of the uncertainty set*. The process
of making an optimization formulation robust can be viewed as a mapping from
one optimization problem to another. A central question is as follows: when will
important properties, such as convexity and decomposability, be preserved under
such mapping? In particular, what kind of nominal formulation and uncertainty set
representation will preserve convexity and decomposability in the robust version of
the optimization problem?

So far, almost all of the work on robust optimization focuses on determining what
representations of data uncertainty preserves convexity, thus tractability through
a centralized solution, in the robust counter part of the nominal problem for a
given definition of robustness. For example, for the worst-case robustness, it has
been shown that under the assumption of ellipsoid set of data uncertainty, a robust
linear optimization problem can be converted into a second-order cone problem;
and a robust second-order cone problem can be reformulated as a semi-definite
optimization problem [26].

In this section, motivated by needs in communication networking, we will focus
instead on the *distributiveness*-preserving formulation of the robust optimization.
The driving question thus becomes: how much more communication overhead is
introduced in making the problem robust?

To develop a systematic theory of distributed robust optimization (DRO), we
first show how to represent an uncertainty set in a way that not only captures
the data uncertainty in the model but also leads to a distributively solvable
optimization problem. Second, in the case where a fully distributed algorithm is
not obtainable, we focus on the *tradeoff* between robustness and distributiveness.
Distributed algorithms are often developed based on decomposability structure of
the problem, which may disappear as the optimization formulation is made robust.
While distributed computation has long been studied, unlike convexity of a problem,
distributiveness of an algorithm does not have a widely agreed definition. It is often

quantified by the amount of communication overhead required: how far and how frequent do the nodes have to pass message around? Zero communication overhead is obviously the "most distributed", and we will see how the amount of overhead trades-off with the degree of robustness.

There are many uncertainties in the design of communication networks. These uncertainties stem from various sources and can be broadly grouped into two categories. The first type of uncertainties are related to the perturbation of a set of design parameters due to erroneous inputs such as errors in estimation or implementation. We call them *perturbation errors*. A common characteristic of such perturbation errors is that they can often be modeled as a *continuous* uncertainty set surrounding the basic point estimate of the parameter. The size of the uncertainty set could be used to characterize the level of perturbations the designer needs to protect against. The second type of uncertainties is termed as *disruptive errors*, as they are caused by the failure of communication links within the network. This type of errors can be modeled as a *discrete* uncertainty set.

As a starter, we will focus on a class of optimization problems with the following nominal form: maximization of a *concave* objective function over a given data set characterized by *linear* constraints,

$$\text{maximize } f_0(x) \tag{2.30}$$

$$\text{subject to } Ax \preceq b$$

$$\text{variables } x,$$

where A is an $M \times N$ matrix, x is an $N \times 1$ vector, and b is an $M \times 1$ vector. This class of problems can model a wide range of engineering systems (e.g., [27]).

The uncertainty of problem (2.30) may exist in the objective function f_0, matrix parameter A, and vector parameter b. In many cases, the uncertainty in objective function f_0 can be converted into uncertainty of the parameters defining the constraints [28]. It is also possible to convert the uncertainty in b into uncertainty in A in certain cases (although this could be difficult in general). In the rest of this section, we will focus on studying the uncertainty in A. In many networking problems, structures and physical meaning of matrix A readily lead to distributed algorithms, e.g., in rate control where A is a given routing matrix, distributed subgradient algorithm is well known to solve the problem, with an interesting correspondence with the practical protocol of TCP. Making the optimization robust may turn the linear constraints into nonlinear ones and increase the amount of message passing.

In the robust counterpart of problem (2.30), we require the constraints $Ax \preceq b$ to be valid for any $A \in \mathcal{A}$, where $\mathcal{A}$ denotes the uncertainty set of A, and the definition of robustness is in the *worst-case* sense [19]. This approach might be too conservative. A more meaningful choice of robustness is the *chance-constrained robustness*, i.e., the probability of infeasibility (or outage) is upper bounded. We can flexibly adjust the chance-constrained robustness of the robust solution by solving the worst-case robust optimization problem over a properly selected subset of the exact uncertainty set.

If we allow an arbitrary uncertainty set A, then the robust optimization problem is difficult to solve even in a centralized manner [29]. We will focus on the study of *constraint-wise* (i.e., *row-wise*) uncertainty set, where the uncertainties between different rows in matrix A are decoupled. This restricted class of uncertainty set characterizes the data uncertainty in many practical problems, and it also allows us to convert the robust optimization problem into a formulation that is distributively solvable.

Denote the jth row of A be a_j^T, which lies in a compact uncertainty set $\mathcal{A}_j$. Then the *robust* optimization problem that we focus on in this chapter can be written in the following form:

$$\text{maximize } f_0(x), \tag{2.31}$$

$$\text{subject to } a_j^T x \leq b_j, \ \forall a_j \in \mathcal{A}_j, \ \forall 1 \leq j \leq M,$$

$$\text{variables } x.$$

It is easy to see that the robust optimization problem (2.31) can be equivalently written in a form represented by *protection functions* instead of uncertainty sets. Denote the nominal counterpart of problem (2.31) with a coefficient matrix $\bar{A}$ (i.e., the values when there is no uncertainty), with the jth row's coefficient $\bar{a}_j \in \mathcal{A}_j$. Then

Theorem 2.6. *Problem (2.31) is equivalent to the following* convex *optimization problem:*

$$\text{maximize } f_0(x), \tag{2.32}$$

$$\text{subject to } \bar{a}_j^T x + g_j(x) \leq b_j, \ \forall 1 \leq j \leq M,$$

$$\text{variables } x,$$

where

$$g_j(x) = \max_{a_j \in \mathcal{A}_j} (a_j - \bar{a}_j)^T x, \tag{2.33}$$

is the protection function for the jth constraint, which depends on the uncertainty set $\mathcal{A}_j$ and the nominal row $\bar{a}_j$. Furthermore, $g_j(x)$ is a convex function since for any $0 \leq t \leq 1$, we have that

$$\max_{a_j \in \mathcal{A}_j} (a_j - \bar{a}_j)^T [tx_1 + (1-t)x_2] \leq t \max_{a_j \in \mathcal{A}_j} (a_j - \bar{a}_j)^T x_1 + (1-t) \max_{a_j \in \mathcal{A}_j} (a_j - \bar{a}_j)^T x_2. \tag{2.34}$$

Different forms of $\mathcal{A}_j$ will lead to different protection function $g_j(x)$, which results in different robustness and performance tradeoff of the formulation. Next we consider several approaches in terms of modeling $\mathcal{A}_j$ and the corresponding protection function $g_j(x)$.

2.4.1.1 Polyhedron Uncertain Set

In this case, the uncertainty set $\mathcal{A}_j$ is a polyhedron characterized by a set of linear inequalities, i.e., $\mathcal{A}_j \triangleq \{a_j : D_j a_j \preceq c_j\}$. The protection function is

$$g_j(x) = \max_{a_j:D_j a_j \preceq c_j} (a_j - \bar{a}_j)^T x, \qquad (2.35)$$

which involves a linear program (LP). We next show that the uncertainty set can be translated into a set of linear constraints. In the jth constraint in (2.31), with $x = \hat{x}$ fixed, we can characterize the set $\forall a_j \in \mathcal{A}_j$ by comparing b_j with the outcome of the following LP:

$$v_j^* = \max_{a_j:D_j a_j \preceq c_j} a_j^T \hat{x}. \qquad (2.36)$$

If $v_j^* \leq b_j$, then $\hat{x}$ is feasible for (2.31). However, this approach is not very useful since it requires solving one LP in (2.36) for each possible $\hat{x}$. Alternatively, we take the Lagrange dual problem of the LP in (2.36),

$$v_j^* = \min_{p_j:D_j^T p_j \succeq \hat{x}, p_j \succeq 0} c_j^T p_j. \qquad (2.37)$$

If we can find a feasible solution $\hat{p}_j$ for (2.37), and $c_j^T \hat{p}_j \leq b_j$, then we must have $v_j^* \leq c_j^T \hat{p}_j \leq b_j$. We can thus replace constraints in (2.31) by the following constraints:

$$c_j^T p_j \leq b_j, \ D_j^T p_j \succeq x, \ p_j \succeq 0, \ \forall 1 \leq j \leq M, \qquad (2.38)$$

and we now have an equivalent and *deterministic* formulation for problem (2.31), where all the constraints are linear.

2.4.1.2 D-Norm Uncertainty Set

D-norm approach [28] is another method to model the uncertainty set, and has advantages such as guarantee of feasibility independent of uncertainty distributions and flexibility in trading off between robustness and performance.

Consider the jth constraint $a_j^T x \leq b_j$ in (2.31). Denote the set of all uncertain coefficients in a_j as $\mathcal{E}_j$. The size of $\mathcal{E}_j$ is $|\mathcal{E}_j|$, which might be smaller than the total number of coefficients N (i.e., a_{ij} for some i might not have uncertainty). For each $a_{ij} \in \mathcal{E}_j$, assume the actual value falls into the range of $[\bar{a}_{ij} - \hat{a}_{ij}, \bar{a}_{ij} + \hat{a}_{ij}]$, in which $\hat{a}_{ij}$ is a given error bound. Also choose a nonnegative integer $\Gamma_j \leq |\mathcal{E}_j|$. The definition of robustness associated with the D-norm formulation is to maintain feasibility if at most Γ_j out of all possible $|\mathcal{E}_j|$ parameters are perturbed. Let us denote $\mathcal{S}_i$ as the set of Γ_j uncertain coefficients. As is well known, the above robustness definition can be characterized by the following protection function,

$$g_j(x) = \max_{S_j : S_j \subseteq \mathcal{E}_j, |S_j| = \Gamma_j} \sum_{i \in S_j} \hat{a}_{ij} |x_i|. \tag{2.39}$$

If $\Gamma_j = 0$, then $g_j(\Gamma_j, x) = 0$ and the jth constraint is reduced to the nominal constraint. If $\Gamma_j = |\mathcal{E}_j|$, then $g_j(x) = \sum_{i \in \mathcal{E}_j} \hat{a}_{ij} |x_i|$, and the jth constraint becomes Soyster's worst-case formulation [28]. The tradeoff between robustness and performance can be obtained by adjusting Γ_j.

2.4.1.3 Ellipsoid Uncertainty Set

Ellipsoid is commonly used to approximate complicated uncertainty sets for statistical reasons [29] and for its ability to succinctly describe a set of discrete points in Euclidean geometry [19]. Here we consider the case where coefficient a_j falls in an ellipsoid centered at the nominal $\bar{a}_j$. Specifically,

$$\mathcal{A}_j = \{\bar{a}_j + \Delta a_j : \sum_i |\Delta a_{ij}|^2 \leq \varepsilon_j^2 \}. \tag{2.40}$$

By (2.33), the protection function is given by

$$g_j(x) = \max \left\{ \sum_i \Delta a_{ij} x_i : \sum_i |\Delta a_{ij}|^2 \leq \varepsilon_j^2 \right\}, \tag{2.41}$$

Denote by $\|x\|_2 = \sqrt{\sum_{i=1}^n x_i^2}$ as the ℓ_2-norm (or the Euclidean norm) of x. By Cauchy–Schwartz inequality,

$$\sum_i \Delta a_{ij} x_i \leq \sqrt{\sum_i |\Delta a_{ij}|^2} \|x\|_2 \leq \varepsilon_j \|x\|_2,$$

and the equality is attained by choosing

$$\Delta a_{ij} = \frac{x_i \varepsilon_j}{\|x\|_2}.$$

Therefore, we conclude that

$$g_j(x) = \varepsilon_j \|x\|_2. \tag{2.42}$$

2.4.2 Robust Multipath Rate Control for Network Service Reliability

2.4.2.1 Nominal and Robust Formulations

Consider a wireline network where some links might fail because of human mistakes, software bugs, hardware defects, or natural hazard. Network operators typically reserve some bandwidth for some backup paths. When the primary paths fail, some or all of the traffic will be rerouted to the corresponding disjoint backup paths. Fast system recovery schemes are essential to ensure service availability in the presence of link failure. There are three key components for fast system recovery [30]: identifying a backup path disjoint from the primary path, computing network resource (such as bandwidth) in reservation prior to link failure, and detecting the link failure in real-time and rerouting the traffic. The first component has been investigated extensively in graph theory. The third component has been extensively studied in system research community. Here we consider the robust rate control and bandwidth reservation in the face of possible failure of primary path, which is related to the second component.

We first consider the nominal problem with no link failures. Following similar notation as in Kelly's seminal work [27], we consider a network with S users, L links, and T paths, indexed by s, l, and t, respectively. Each user is a unique flow from one source node to one destination node. There could be multiple users between the same source–destination node pair. The network is characterized by the $L \times T$ path-availability $0 - 1$ matrix

$$[D]_{lt} = \begin{cases} d_{lt} = 1, & \text{if link } l \text{ is on path } t, \\ 0, & \text{otherwise.} \end{cases}$$

and $T \times S$ primary-path-choice nonnegative matrix

$$[W]_{ts} = \begin{cases} w_{ts}, & \text{if user } s \text{ uses path } t \text{ as the primary path,} \\ 0, & \text{otherwise.} \end{cases}$$

where $w_{ts} > 0$ indicates the percentage that user s allocates its rate to primary path t, and $\sum_t w_{ts} = 1$. Let x, c, and y denote source rates, link capacities, and aggregated path rates, respectively. The nominal multipath rate control problem is

$$\text{maximize} \quad \sum_s f_s(x_s) \tag{2.43}$$

$$\text{subject to } Dy \preceq c, \quad Wx \preceq y,$$

$$\text{variables } x \succeq 0, y \succeq 0,$$

where $f_s(x_s)$ is user s' utility function, which is increasing and strictly concave in x_s.

To represent Problem (2.43) in a more compact way, we denote $R = DW$ as the link-source matrix

$$[R]_{ls} = \begin{cases} r_{ls} = \sum_{t \in T(l)} w_{ts}, & \text{if user } s \text{ uses link } l \text{ in one of its primary paths,} \\ 0, & \text{otherwise.} \end{cases}$$

where $T(l)$ denotes the set of all paths associated with link l, i.e., $T(l) = \{t : d_{lt} = 1\}$. The nominal problem can be compactly rewritten as

$$\text{maximize } \sum_s f_s(x_s) \tag{2.44}$$

$$\text{subject to } Rx \preceq c,$$

$$\text{variables } x \succeq 0.$$

To ensure robust data transmission against the link failures, each user also determines a backup path when it joins the network. The nonnegative backup path choice matrix is

$$[B]_{ts} = \begin{cases} b_{ts}, & \text{if user } s \text{ uses path } t \text{ as the backup path,} \\ 0, & \text{otherwise.} \end{cases}$$

where $b_{ts} > 0$ indicates the maximum percentage that user s allocates its rate to path t. The actual rate allocation will be a random variable between 0 and b_{ts}, depending on whether the primary paths fail. We further assume that a path can only be used as either a primary path or a backup path for the same user but not both. The corresponding *robust multipath routing rate allocation problem* is given by

$$\text{maximize } \sum_s f_s(x_s) \tag{2.45}$$

$$\text{subject to } \sum_{s \in S(l)} r_{ls} x_s + \sum_{t \in T(l)} g_t(b_t, x) \leq c_l, \ \forall l.$$

$$\text{variables } x \succeq 0.$$

Here $S(l)$ denotes the set of users using the link l in one of their primary paths, and $\sum_{s \in S(l)} r_{ls} x_s$ is the aggregate rate from users. Moveover, $g_t(b_t, x)$ is the protection function for the traffic from users who use path t as their backup path, and b_t is the tth row of matrix B.

There are several ways of characterizing the protection function. Here we take D-norm as an example. Let $\mathcal{E}_t = \{s : b_{ts} > 0, \forall s\}$ denote the set of users who utilize path t as the backup path, and $\mathcal{F}_{t,\Gamma_t}$ denote a set such that

$$\mathcal{F}_{t,\Gamma_t} \subseteq \mathcal{E}_t \text{ and } |\mathcal{F}_{t,\Gamma_t}| = \Gamma_t.$$

Here Γ_t denotes the number of users who might experience path failures, and its value controls the tradeoff between robustness and performance. The protection function can then be written as

$$g_t(b_t, x) = \max_{\mathcal{F}_{t,\Gamma_t} \subseteq \mathcal{E}_t} \sum_{s \in \mathcal{F}_{t,\Gamma_t}} b_{ts} x_s, \forall t. \qquad (2.46)$$

Notice that Γ_t is a parameter (instead of a variable) in the protection function (2.46).

A centralized algorithm such as the interior point method can be used to solve the service reliability problem (2.45). In practice, however, a distributed solution is preferred in many situations. For example, for a large-scale communication network, multiple nodes and associated links might be removed or updated, and it could be difficult to collect the updated topology information of the whole network. Furthermore, a distributed algorithm can be used to dynamically adjust the rates according to changes in the network topology. Since we consider service reliability in this application, most likely such a update is only required infrequently.

2.4.2.2 Distributed Primal–Dual Algorithms

We now develop a fast distributed algorithm to solve the robust optimization of multipath rate control based on a combination of active-set method [31] and dual-based decomposition method.

We first show that the nonlinear constraints in Problem (2.45) can be replaced by a set of linear constraints:

Proposition 2.1. *For any path t, the single constraint*

$$\sum_{s \in S(l)} r_{ls} x_s + \sum_{t \in T(l)} g_t(b_t, x) \leq c_l, \qquad (2.47)$$

is equivalent to the following set of constraints

$$\sum_{s \in S(l)} r_{ls} x_s + \sum_{t \in T(l)} \sum_{s \in \mathcal{F}_{t,\Gamma_t}} b_{ts} x_s \leq c_l, \forall (\mathcal{F}_{t,\Gamma_t}, \forall t \in T(l)) \text{ such that } \mathcal{F}_{t,\Gamma_t} \subseteq \mathcal{E}_t. \quad (2.48)$$

We further define some short-hand notation. For each choice of parameters ($|T(l)|$ sets of users)

$$\mathcal{F}_l = (\mathcal{F}_{t,\Gamma_t} \subseteq \mathcal{E}_t, \forall t \in T(l)),$$

we define a set corresponding to the choices of paths and users:

$$\mathcal{Q}_{\mathcal{F}_l} \triangleq \{(t, s) : t \in T(l), s \in \mathcal{F}_{t,\Gamma_t}\}.$$

We further define $\mathcal{G}_l = \{\mathcal{Q}_{\mathcal{F}_l}, \forall \mathcal{F}_l\}$. The constraints in (2.48) can be rewritten as follows:

$$\sum_{s \in S(l)} r_{ls} x_s + \sum_{(t,s) \in \mathcal{Q}_{\mathcal{F}_l}} b_{ts} x_s \leq c_l, \ \forall \mathcal{Q}_{\mathcal{F}_l} \in \mathcal{G}_l. \tag{2.49}$$

Note the number of constraints in (2.49) is $\prod_{t \in T(l)} \binom{|\mathcal{E}_t|}{\Gamma_t}$, and increases quickly with Γ_t and $|\mathcal{E}_t|$. This motivates us to design an alternative method to solve (2.45). We next present a sequential optimization approach. The basic idea is to iteratively generate a set $\bar{\mathcal{G}}_l \subseteq \mathcal{G}_l$, and use the following set of constraints to approximate (2.49):

$$\sum_{s \in S(l)} r_{ls} x_s + \sum_{(t,s) \in \mathcal{Q}_{\mathcal{F}_l}} b_{ts} x_s \leq c_l, \ \forall \mathcal{Q}_{\mathcal{F}_l} \in \bar{\mathcal{G}}_l. \tag{2.50}$$

This leads to a relaxation of Problem (2.45):

$$\text{maximize} \ \sum_s f_s(x_s) \tag{2.51}$$

$$\text{subject to} \ \sum_{s \in S(l)} r_{ls} x_s + \sum_{(t,s) \in \mathcal{Q}_{\mathcal{F}_l}} b_{ts} x_s \leq c_l, \ \forall \mathcal{Q}_{\mathcal{F}_l} \in \bar{\mathcal{G}}_l, \ \forall l,$$

$$\text{variables} \ x \succeq 0 \,.$$

Let $\bar{x}$ denote an optimal solution to the relaxed problem (2.51) and x^* denote an optimal solution of the original problem (2.45). If $\bar{\mathcal{G}}_l = \mathcal{G}_l$, then we have $\sum_s f_s(\bar{x}_s) = \sum_s f_s(x_s^*)$. Even if $\bar{\mathcal{G}}_l \subset \mathcal{G}_l$, it is still possible that the two optimal objective values are the same.

Proposition 2.2. $\sum_s f_s(\bar{x}_s) = \sum_s f_s(x_s^*)$ *if the following condition holds*

$$\max_{\mathcal{Q}_{\mathcal{F}_l} \in \bar{\mathcal{G}}_l} \sum_{(t,s) \in \mathcal{Q}_{\mathcal{F}_l}} b_{ts} \bar{x}_s = \max_{\mathcal{Q}_{\mathcal{F}_l} \in \mathcal{G}_l} \sum_{(t,s) \in \mathcal{Q}_{\mathcal{F}_l}} b_{ts} \bar{x}_s, \ \forall l. \tag{2.52}$$

Next we develop a distributed algorithm to solve Problem (2.51) for a fixed $\bar{\mathcal{G}}_l$ for each l, which is suboptimal for solving Problem (2.45). We can then design an optimal distributed algorithm that achieves the optimal solution of Problem (2.45) by iteratively using the following algorithm.

First, by relaxing the constraints in Problem (2.51) using dual variables $\lambda = \{\{\lambda_{li}\}_{i=1}^{|\bar{\mathcal{G}}_l|}\}_{l=1}^{L}$, we obtain the following Lagrangian,

$$\bar{Z}(\lambda, x) = \sum_s f_s(x_s) + \sum_l \sum_{i=1}^{|\bar{\mathcal{G}}_l|} \lambda_{li} \left(c_l - \sum_{s \in S(l)} r_{ls} x_s - \sum_{(t,s) \in \mathcal{Q}_{\mathcal{F}_l}(i)} b_{ts} x_s \right).$$

For easy indexing, here we denote each set $\bar{\mathcal{G}}_l = \{\mathcal{Q}_{\mathcal{F}_l}(i), i = 1, \dots, |\bar{\mathcal{G}}_l|\}$.

Distributed Primal-dual Subgradient Algorithm

$k \leftarrow 0, \boldsymbol{\lambda}(0) \leftarrow \mathbf{0}, \boldsymbol{\mu}(0) \leftarrow \mathbf{0}$, and choose $\epsilon \ll 1$

while there exists an i so that $\eta_i(k) = 1 - \dfrac{\sum_s f_s(\bar{x}_{s,i}(k))}{Z(\lambda_k)} > \epsilon_i$

    $k \leftarrow k+1$

    Each user s determines $x_s(k)$ by solving Problem (1.54)

    Each user s passes the value of $x_s(k)$ to each link associated with this user

    Each user s who uses link l on either its primary or backup paths calculates

       the associated dual prices by passing messages over path t

       from its destination to its source

    Each user s computes $\bar{\Delta}_{s,i}(k) = \min_l \left[\left(\bar{\Delta}_s(l,k) \right)_{r_{ls}>0}, \left(\bar{\Delta}_s(l,k) \right)_{b_{ts}>0,\ (t,s) \in \mathcal{Q}_{\mathcal{F}_l}(i)} \right]$,

    and $\bar{x}_{s,i}(k) = x_s(k) \bar{\Delta}_{s,i}(k)$ by collecting messages from its associated links

end while

Fig. 2.5 Distributed primal–dual subgradient algorithm for robust optimization of multipath rate control

The dual function is

$$Z(\lambda) = \max_{x \succeq 0} \bar{Z}(\lambda, x). \tag{2.53}$$

The optimization over x in (2.53) can be decomposed into one problem for each user s:

$$\max_{x_s \geq 0} \left(f_s(x_s) - \sum_l \sum_{i=1}^{|\bar{\mathcal{G}}_l|} \left(\lambda_{li} r_{ls} + \lambda_{li} \sum_{(t,s) \in \mathcal{Q}_{\mathcal{F}_l}(i)} b_{ts} \right) x_s \right). \tag{2.54}$$

Notice that link l can be viewed as the composition of $|\bar{\mathcal{G}}_l|$ sub-links and each sub-link is associated with a price (dual variable) λ_{li}. Each user s determines its transmission rate x_s by considering prices from both its primary path and backup path.

The master dual problem is

$$\min_{\lambda \succeq 0} Z(\lambda), \tag{2.55}$$

which can be solved by the subgradient method. For each dual variable λ_{li}, its subgradient can be calculated as

$$\zeta_{li}(\lambda_{li}) = c_l - \sum_{s \in S(l)} r_{ls} x_s - \sum_{(t,s) \in \mathcal{Q}_{\mathcal{F}_l}(i)} b_{ts} x_s. \tag{2.56}$$

The value of λ_{li} will be updated using the subgradient information correspondingly.

Pseudo-code of the algorithm is in Fig. 2.5. Here $\theta(k)$ is the step-size at time k. If it meets conditions such as those given in [32, p. 505], then this subgradient method is guaranteed to converge to the optimal solution. Common choices of step size include the constant step size ($\theta(k) = \beta$), diminish step size ($\theta(k) = \frac{\beta}{\sqrt{k}}$), and square summable ($\theta(k) = \frac{\beta}{k}$). The parameter ε_i is a predefined small threshold.

Due to the special structure of (2.51), we can characterize the suboptimality gap of the obtained solution at each iteration, as shown below.

Proposition 2.3. *Let $P^*(\{\bar{\mathcal{G}}_l\})$ be the optimal object function value of the relaxed problem (2.51), λ^k be the dual variables at the kth iteration of the algorithm above, and $[\bar{x}_1(k),\ldots,\bar{x}_S(k)]$ be a feasible solution to (2.51). Then $P^*(\{\bar{\mathcal{G}}_l\})$ is lower bounded by $\sum_s f_s(\bar{x}_s(k))$ and upper bounded by $Z(\lambda^k)$, i.e.,*

$$\sum_s f_s(\bar{x}_s(k)) \leq P^*(\{\bar{\mathcal{G}}_l\}) \leq Z(\lambda^k). \tag{2.57}$$

Furthermore, the suboptimality gap $\eta_i(k)$ satisfies $\lim_{k\to\infty} \eta_i(k) = 0$ for all i.

2.5 Summary

To summarize, these applications to rate control and P2P capacity are only small samples of the use of distributed optimization in networking. This chapter serves as a terse teaser. Much more on existing results, including proofs and numerical examples, can be found from the papers referenced here. Much more also remains to be explored as the research community continues to formulate difficult and robust optimization problems and try out distributed solutions.

Acknowledgements Minghua Chen was partially supported by the China 973 Program (Project No. 2012CB315904), and the General Research Fund grants (Project Nos. 411209, 411010, and 411011) and an Area of Excellence Grant (Project No. AoE/E-02/08), established under the University Grant Committee of the Hong Kong SAR, China, as well as two gift grants from Microsoft and Cisco. Mung Chiang was partially supported by AFOSR MURI grant FA9550-09-1-0643.

References

1. Sengupta, S., Liu, S., Chen, M., Chiang, M., Li, J., Chou, P.: Peer-to-peer streaming capacity. IEEE Trans. Inform. Theory **57**(8), 5072–5087 (2011)
2. Liu, S., Chen, M., Sengupta, S., Chiang, M., Li, J., Chou, P.: "Peer-to-Peer Streaming Capacity under Node Degree Bound", Proceedings of the 30th International Conference on Distributed Computing Systems (ICDCS 2010), Genoa, Italy, 21–25 (2010)
3. Zhang, S., Shao, Z., Chen, M.: Optimal distributed p2p streaming under node degree bounds. In: Proceedings of the IEEE ICNP, Kyoto, Japan (2010)
4. Garg, N., Konemann, J.: Faster and simpler algorithms for multicommodity flow and other fractional packing problems. Proceedings of the 39th IEEE Symposium on Foundations of Computer Science (FOCS 1998), Palo Alto, CA, USA, 8–11 (1998)

5. Dantzig, G., Wolfe, P.: Decomposition principle for linear programs. Oper. Res. **8**, 101–111 (1960)
6. Gilmore, P., Gomory, R.: A linear programming approach to the cutting-stock problem. Oper. Res. **9**, 849–859 (1961)
7. Gilmore, P., Gomory, R.: A linear programming approach to the cutting stock problem—Part II. Oper. Res. **11**, 863–888 (1963)
8. Lin, X., Shroff, N., Srikant, R.: A tutorial on cross-layer optimization in wireless networks. IEEE J. Sel. Areas Commun. **24**(8), 1452–1463 (2006)
9. Jiang, L., Walrand, J.: "A distributed csma algorithm for throughput and utility maximization in wireless networks". In: Proceedings of the 46th Annual Allerton Conference on Communication, Control, and Computing, Urbana, IL, USA (2008)
10. Chen, M., Liew, S., Shao, Z., Kai, C.: Markov approximation for combinatorial network optimization. CUHK Technical Report (2009). Available at http://www.ie.cuhk.edu/~mhchen/ma.tr.pdf
11. Zhang, H., Chen, M., Parekh, A., Ramchandran, K.: An adaptive multi-channel p2p video-on-demand system using plug-and-play helpers. EECS Department, University of California, Berkeley, Tech. Rep. UCB/EECS-2010-111, July 2010. [Online] Available: http://www.eecs.berkeley.edu/Pubs/TechRpts/2010/EECS-2010-111.html
12. Jiang, W., Lan, T., Ha, S., Chen, M., Chiang, M.: "Joint VM Placement and Routing for Data Center Traffic Engineering". In: Proceedings of IEEE INFOCOM (mini-conference), Orlando, Florida, USA, 25–30 (2012)
13. Chen, M., Liew, S.C., Shao, Z., Kai, C.: Markov approximation for combinatorial network optimization. In: Proceedings of the IEEE INFOCOM, San Diego (2010)
14. Meng, X., Pappas, V., Zhang, L.: Improving the scalability of data center networks with traffic aware virtual machine placement. In: Proceedings of the IEEE INFOCOM 2010, San Diego, CA, US (2010)
15. Garg, N., Konjevod, G., Ravi, R.: A polylogarithmic approximation algorithm for the group Steiner tree problem. In the 9th Annual ACM-SIAM SODA (1998), San Francisco, CA, USA, 25–27 (1998)
16. Feige, U.: A threshold of ln n for approximating set cover. In: Proceedings of the 28th ACM STOC (1996), Philadelphia, PA, USA, 22–24 (1996)
17. Charikar, M., Chekuri, C., Cheung, T., Dai, Z., Goel, A., Guha, S., Li, M.: Approximation algorithms for directed Steiner tree problems. In: Proceedings of the 9th Annual ACM-SIAM SODA (1998), San Francisco, CA, USA, 25–27 (1998)
18. Chiang, M.: Geometric programming for communication systems, Now Publishers Inc (2005)
19. Boyd, S., Vandenberghe, L.: Convex Optimization. Cambridge University Press, Cambridge, UK (2004)
20. Diaconis, P., Stroock, D.: Geometric bounds for eigenvalues of Markov chains. Ann. Appl. Prob. **1**, 36–61 (1991)
21. Ben-Tal, A., Nemirovski, A.: Robust solutions to uncertain programs. Oper. Res. Lett. **25**(1), 1–13 (1999)
22. Ben-Tal, A., Nemirovski, A.: Selected topics in robust convex optimization. Math. Program. **112**(1), 125–158 (2008)
23. Nemirovski, A.: On tractable approximations of randomly perturbed convex constraints. In Proceedings of the 42nd IEEE Conference on Decision and Control, Maui, HI, USA, 8–12 (2003)
24. El-Ghaoui, L.: Robust solutions to least-square problems to uncertain data matrices. SIAM J. Matrix Anal. Appl. **18**, 1035–1064 (1997)
25. El Ghaoui, L., Oustry, F., Lebret, H.: Robust solutions to uncertain semidefinite programs. SIAM J. Optim. **9**, 33 (1998)
26. Ben-Tal, A., Nemirovski, A.: Robust solutions to linear programming problems contaminated with uncertain data. Math. Program. **88**, 411–424 (2000)
27. Kelly, F., Maulloo, A., Tan, D.: Rate control for communication networks: shadow prices, proportional fairness and stability. J. Oper. Res. Soc. 237–252 (1998).

28. Bertsimas, D., Sim, M.: The price of robustness. Oper. Res. **52**(1), 35–53 (2004)
29. Ben-Tal, A., Nemirovski, A.: Robust solutions of uncertain linear programs. Oper. Res. Lett. **25**(1), 1–14 (1999)
30. Xu, D., Li, Y., Chiang, M., Calderbank, A.: Optimal provision of elastic service availability. In Proceedings of IEEE INFOCOM (2007), Anchorage, AK, USA, 6–12 (2007)
31. Leyffer, S.: The return of the active set method. Oberwolfach Rep. **2**(1) (2005)
32. Bertsimas, D., Tsitsiklis, J.N.: Introduction to Linear Optimization. Athena Scientific, Belmont, Mass., USA (1997)

Chapter 3
Fast First-Order Algorithms for Packing–Covering Semidefinite Programs

G. Iyengar, D.J. Phillips, and C. Stein

3.1 Introduction

Semidefinite programming (SDP) is used to solve a wide variety of computational problems. In recent years, it has been used to find approximate solutions for NP-hard optimization problems such as Vertex Cover [12], MIN UNCUT[1] and unrelated machine scheduling [18]. For these problems, one first solves an SDP relaxation of the original problem and then employs some type of rounding algorithm in order to find a solution to the original problem. Other applications of SDP include synchronous code division multiple access (CDMA) [19] and sensor localization [5], although approximation algorithms are not used in these applications.

For all these problems, the computational bottleneck is the solution of the SDP. Both in terms of theoretical worst case and practical running time, solving large SDPs is significantly more difficult than solving large linear programs. In this chapter we focus on algorithms which find ε-optimal solutions in time polynomial in $1/\varepsilon$ and in the size of the SDP. Thus, if one can write a polynomial sized SDP, one then gets an approximation algorithm for the application at hand. Within this realm there have been at least four lines of work on finding solutions for SDPs.

One approach is to use specialized interior point methods. These run in time polynomial in $\log(1/\varepsilon)$ but tend to have large dependence on the size of the SDP [16]. Because of the slow running times, there has been work on finding solutions to special classes of SDPs that arise in applications, and giving algorithms with higher dependence on ε but better dependence on the other parameters.

G. Iyengar • C. Stein
The Department of Industrial Engineering & Operations Research, Columbia University, New York, NY, USA
e-mail: garud@ieor.columbia.edu; cliff@ieor.columbia.edu

D.J. Phillips (✉)
Mathematics Department, United States Naval Academy, Annapolis, MD, USA
e-mail: dphillip@usna.edu

T. Terlaky and F.E. Curtis (eds.), *Modeling and Optimization: Theory and Applications*, Springer Proceedings in Mathematics & Statistics 21, DOI 10.1007/978-1-4614-3924-0_3, © Springer Science+Business Media, LLC 2012

The other three approaches all focus on solving special classes of SDPs. The first such approach by Klein and Lu [11] generalizes ideas from solving packing linear programs [17] to give algorithms whose dependence on epsilon is $\Omega(\varepsilon^{-2})$ but tend to be very efficient in terms of the other parameters. Their algorithms were only applicable to a few applications, including MAX CUT and coloring. A third, related approach by Arora and Kale [2] tends to have even worse dependence on ε while achieving very good dependence on the other parameters. A notable feature of [2] is that their algorithms are specifically for SDP relaxations for combinatorial NP-Hard problems and guarantee primal feasibility or a direct rounding to the underlying NP-Hard problem.

A fourth approach obtains running times with a $1/\varepsilon$ dependence, while typically realizing a slower dependence on the other parameters than the two previously mentioned approaches. This approach, introduced by Iyengar et al. [9, 10], identifies classes of SDPs that have a structure that can be exploited using techniques that generalize those used for approximating packing and covering linear programs [17] and build upon work of Nesterov [14] for approximating more general classes of problems. In this line of work the authors have identified two classes of SDPs, which they call *packing SDPs* and *covering SDPs*, due to their relation to packing and covering linear programs (see [17]). For each of these classes, they give efficient algorithms with $\mathcal{O}(\varepsilon^{-1})$ dependence and which capture classes of SDPs that arise in applications. The distinguishing feature of packing (or covering) SDPs are the presence of inequality constraints, $\mathbf{Tr}(A^\top X) \leq 1$ ($\mathbf{Tr}(A^\top X) \geq 1$) for a given parameter matrix $A \succeq 0$ and decision variable matrix X constrained to be positive semidefinite. Hence, packing constraints are generalized knapsack constraints and covering constraints are generalized covering constraints. Packing and covering algorithms typically use Lagrangian penalization to solve these problems. Practical experience with these types of problems [4, 8] shows that running time dependence on ε is a key factor in designing an efficient, practical implementation. A clear drawback of the packing [9] and covering [10] algorithms is that they cannot efficiently approximate SDPs that have *both* packing and covering constraints, and, in particular, these algorithms cannot efficiently approximate SDPs with equality constraints. Equality constraints arise in many applications including the SDPs used for Vertex Color, MIN UNCUT, and unrelated machine scheduling. In some applications the equality constraints can be removed by appropriately reformulating the problem; however, in many others there are no obvious methods for removing equality constraints. For example, in the Vertex Cover SDP the equality constraints are of the form $\mathbf{diag}(X) = \mathbf{1}$. Replacing them by *two* inequality constraints—$\mathbf{diag}(X) \geq \mathbf{1}$ and $\mathbf{diag}(X) \leq \mathbf{1}$—does not preserve either the packing or the covering property. Only imposing $\mathbf{diag}(X) \leq \mathbf{1}$ yields a problem with a trivial optimal solution infeasible to the original problem; whereas only imposing $\mathbf{diag}(X) \geq \mathbf{1}$ yields an optimal solution infeasible to the original problem. For the latter, scaling or shifting (as done in [9, 10]) does not preserve positive semidefiniteness.

In order to use the ideas from the packing and covering methods, we identify structure in the equality constraints and SDPs, namely that the SDPs must have strictly feasible solution and that polynomially separated bounds exist on the optimal objective value (see Definition 3.2). Our contributions in this chapter are as follows:

(a) We propose a first-order algorithm that computes a δ-feasible and ε-optimal solution to a packing–covering SDP, i.e., an SDP with both packing [9] and covering [10] constraints. (See Definition 3.1, Sect. 3.2 for the precise definition of δ-feasible solutions.) Our algorithm approximately solves a packing–covering SDP by solving a sequence of relaxed packing–covering SDPs. Each of these relaxed SDPs are solved using a Lagrangian relaxation.

(b) We show how our proposed method can be applied to solve the SDPs that arise in the context of Vertex Cover, MIN UNCUT, and minimizing weighted completion time on two unrelated machines. In all three problems, there are side constraints that can significantly improve the quality of the SDP relaxation. In the case of Vertex Cover, there are $\mathcal{O}(n^2)$ such constraints which means that interior point methods require $\tilde{\mathcal{O}}(n^{6.5})$ time (log factors are suppressed) to solve the associated SDP. The number of additional constraints in the SDPs used for the MIN UNCUT and the Skutella scheduling algorithms are $\mathcal{O}(n^3)$ and $\mathcal{O}(n^2)$, respectively. Thus, interior point is not practical for solving large instances of these problems. For all three packing–covering SDPs, our method runs in $\tilde{\mathcal{O}}(\varepsilon^{-2} \cdot n^4)$ time where we have set the feasibility error to $\delta = \varepsilon$. Kale and Arora have an algorithm for the MIN UNCUT SDP that requires $\tilde{\mathcal{O}}(\min\{nm^2 \cdot \varepsilon^{-4}, n^3 \cdot \varepsilon^{-6}\})$ as opposed to the $\tilde{\mathcal{O}}(n^{9.5}\log(\varepsilon^{-1}))$ running time for interior point methods. So the running time for our method for MIN UNCUT is in between these two extremes in terms of the tradeoff between running time as a function of n and m and running time as a function of ε^{-1}.

3.1.1   Preliminaries

Consider an optimization problem $v^* = \min_{X \in S} f(X)$ where $S \subset \mathbb{R}^n$ such that $v^* > 0$. For $\varepsilon > 0$, we say $Y \in S$ is ε-absolute-optimal if $f(Y) \leq v^* + \varepsilon$, and ε-relative-optimal if $f(Y) \leq (1+\varepsilon)v^*$. We use the notation $\tilde{\mathcal{O}}(f(n, \frac{1}{\varepsilon}))$ to mask logarithmic factors in n and $\frac{1}{\varepsilon}$. We use $\ell(n, m)$ to denote the running required to solve a linear program with $\mathcal{O}(n)$ variables and $\mathcal{O}(m)$ constraints. For matrices $A, B \in \mathbb{R}^{n \times n}$, we define $\langle A, B \rangle = \mathbf{Tr}(A^\top B) = \sum_{ij} A_{ij} B_{ij}$, which is the *Frobenius inner product*. We denote the set of symmetric $n \times n$ matrices by $\mathcal{S}^n$.

3.2 The Packing–Covering SDP

Consider the SDP

$$
\begin{aligned}
\theta^* = \min \ & \langle C, X \rangle \\
\text{s.t.} \quad & \langle A_i, X \rangle \geq 1 & i = 1, \ldots, m_1 \\
& \langle B_j, X \rangle = 1 & j = 1, \ldots, m_2 \\
& X \in \mathcal{X} = \{ Y : \mathbf{Tr}(Y) \leq \tau \},
\end{aligned} \tag{3.1}
$$

where $C, A_i, B_j \in \mathcal{S}^n$ for $i = 1, \ldots, m_1$, and $j = 1, \ldots, m_2$. Unlike in [9, 10] we do *not* require that A_i be semidefinite, in particular, we allow indefinite A_i. We define $m = m_1 + 2m_2$ to be the total number of inequality constraints (since each equality constraint corresponds to two inequalities). Note that both packing and covering constraints can be formulated in the form $\langle A_i, X \rangle \geq 1$ by a suitable scaling and shifting. We define ε-optimal, δ-feasible solution to (3.1) as follows.

Definition 3.1. $Y \in \mathcal{X}$ is an ε-optimal δ-feasible solution for (3.1) if $\langle A_i, Y \rangle \geq 1$, $i = 1, \ldots, m_1$, $|\langle B_j, Y \rangle - 1| < \delta$, $j = 1, \ldots, m_2$, and $\langle C, Y \rangle \leq (1 + \varepsilon)\theta^*$.

As noted in the introduction, to use packing and covering methods, the SDPs must have structure in the form of a constraint condition as well as bounds on the objective. We make this structure clear in the next definition.

Definition 3.2. An SDP of the form (3.1) is called a *packing–covering* SDP if there exists $\overline{Y} \in \mathcal{X}$ that is feasible for (3.1), $\delta_0 \in (0, 1)$, $\theta_L > 0$ and an objective bounding function $p(n, m)$ such that

(i) $0 < \theta_L \leq \min \left\{ \langle C, X \rangle : \langle A_i, X \rangle \geq 1 - \delta_0, \forall\, i, |\langle B_j, X \rangle - 1| \leq \delta_0, \forall\, j, X \in \mathcal{X} \right\}$

(ii) $\dfrac{\langle C, \overline{Y} \rangle}{\theta_L} \leq p(n, m)$

The polynomial separation in the bounds guarantees that the bisection used in our algorithm terminates in a polynomial number of steps. In all of our applications, the bounding function $p(n, m)$ is either constant or logarithmic in n and m.

3.3 Approximating the Packing–Covering SDP

In this section we describe how to find an ε-optimal, δ-feasible solution to packing–covering SDP. Our overall solution approach is as follows. Given a feasibility tolerance $\delta \in (0, \delta_0)$ we first construct the relaxed SDP (3.2). Starting with the initial solution $\overline{Y}$ that is always *strictly* feasible for the relaxed packing–covering SDP (3.2), we use Lagrangian relaxation to construct a sequence of solutions with progressively improving objective function values. The core of the algorithm is described in Sect. 3.3.1. Given $\alpha \in (0, 1)$, we describe how we construct an $\alpha\sigma$-absolute-optimal *strictly* feasible solution for (3.2) using Lagrangian relaxation and σ-absolute-optimal *strictly* feasible solution Z. In Sect. 3.3.2, we describe a bisection procedure that recursively calls the Lagrangian relaxation subroutine to improve optimality guarantees.

3.3.1 Approximating the Relaxed Packing–Covering SDP

Fix $\delta \in (0, \delta_0)$ and define the relaxed packing–covering SDP

$$
\begin{aligned}
v_\delta := \min \quad & \langle C, X \rangle \\
\text{s.t.} \quad & \langle A_i, X \rangle \geq 1 - \delta \; \forall i \\
& \langle B_j, X \rangle \geq 1 - \delta \; \forall j \\
& -\langle B_j, X \rangle \geq -1 - \delta \; \forall j \\
& X \in \mathcal{X}.
\end{aligned}
\tag{3.2}
$$

Note that $v_\delta \geq \theta_L > 0$ for all $\delta < \delta_0$. We abuse notation and refer to both the problem (3.2) and the optimal objective value as v_δ. In this section our goal is to describe a procedure that takes as input a parameter $\alpha \in (0, 1)$ and a σ-optimal *strictly* feasible solution Z and returns a $\alpha\sigma$-optimal *strictly* feasible solution $\hat{Z}$.

For notational convenience, define

$$
(M_i, b_i) = \begin{cases}
(A_i, 1), \; i = 1, \dots, m_1, \\
(B_i, 1), \; i = m_1 + 1, \dots, m_1 + m_2, \\
(-B_i, -1) \; i = m_1 + m_2 + 1, \dots, m_1 + 2m_2,
\end{cases}
$$

Let $\mathcal{V} = \{v \geq 0 : \sum_{i=1}^{m} v_i \leq 1\}$ and fix $\gamma > 0$. Define a Lagrangian saddle-point problem as follows:

$$
v_\delta(\gamma) = \min_{X \in \mathcal{X}} \max_{v \in \mathcal{V}} \left\{ \langle C, X \rangle + \gamma \sum_{i=1}^{m} (b_i - \langle M_i, X \rangle - \delta) v_i \right\}.
$$

Lagrangian duality theory implies that $\lim_{\gamma \to \infty} v_\delta(\gamma) = v_\delta$. For fixed γ and $\varepsilon > 0$, a primal–dual pair $(Z, u) \in \mathcal{X} \times \mathcal{V}$ is said to be a ε-approximate saddle point to $v_\delta(\gamma)$ if

$$
\max_{v \in \mathcal{V}} \left\{ \langle C, Z \rangle + \gamma \sum_{i=1}^{m} (b_i - \langle M_i, Z \rangle - \delta) v_i \right\} - \min_{X \in \mathcal{X}} \left\{ \langle C, X \rangle + \gamma \sum_{i=1}^{m} (b_i - \langle M_i, X \rangle - \delta) u_i \right\} < \varepsilon.
$$

Define the "smoothed" dual function $f_\alpha(v, \delta)$ as follows:

$$
f_\alpha(v, \delta) = \gamma \sum_{i=1}^{m} (b_i - \delta) v_i + \min_{X \in \mathcal{X}} \left\{ \left\langle C - \gamma \sum_{i=1}^{m} v_i M_i, X \right\rangle + \alpha ||X||_F^2 \right\},
$$

where $||X||_F = \sqrt{\langle X, X \rangle}$ denotes the *Frobenius norm* of X. We adapt Nesterov's saddle-point algorithm [14] to find a relatively approximate solution to the problem of minimizing f_α over the set $\mathcal{X}$. In order to use the Nesterov saddle-point algorithm [14], the calculation of the gradient $\nabla_v f(v, \delta)$ is required, which reduces to solving a convex quadratic program. We can use the subroutine in n[10] to find the gradient $\nabla_v f(v, \delta)$ in $\mathcal{O}(n \log n)$ time (see Lemma 2 of [10]). Then the Nesterov saddle-point algorithm [14, 15] (see also [10]) has the following running time.

Theorem 3.1 ([14]). *Let $\varepsilon > 0$, $\delta \in (0, \delta_0)$, and $\gamma > 0$ be given. The Nesterov saddle-point algorithm computes an ε-approximate saddle point to $v_\delta(\gamma)$ in $O\left(\frac{\gamma}{\varepsilon} \cdot \tau \cdot ||M|| \cdot n \ln(n) \cdot \sqrt{\ln(m)}\right)$ time, where $||M|| = \max_i \{\lambda_{\max}(M_i)\}$.*

Note that the running time of the Nesterov algorithm increases with the penalty parameter γ. Next, we show how to "round" a saddle-point to a feasible solution for (3.2). Define the function $g_\delta : \mathcal{X} \to \mathbb{R}$ as follows

$$g_\delta(Z) = \min_{i=1,\ldots,m} \left\{ \langle M_i, Z \rangle - b_i + \delta \right\}.$$

Note that $g_\delta(Z) < 0$ indicates that Z is infeasible to the problem v_δ and $g_\delta(Z) > 0$ indicates Z is *strictly feasible* to v_δ. We let

$$g_\delta^-(Z) = \max\{0, -g(Z)\}$$

denote the absolute infeasibility for a given $Z \in \mathcal{X}$. The following result is established in [10] (see also [13]) .

Lemma 3.1 ([10]). *Fix $\delta \in (0, \delta_0)$ and $\sigma > 0$. Suppose Z is strictly feasible for v_δ and let $\hat{X}$ be a σ-approximate saddle point to $v_\delta(\gamma)$ for $\gamma \geq \frac{\langle C, Z \rangle - v_\delta}{g_\delta(Z)} \geq \frac{\langle C, Z \rangle - \theta_L}{g(Z)}$. Then*

$$\bar{X} = \frac{g_\delta(Z)\hat{X} + g_\delta^-(\hat{X})Z}{g_\delta(Z) + g_\delta^-(\hat{X})}$$

is feasible and σ-absolute-optimal solution for v_δ.

Lemma 3.1 and Theorem 3.1 together imply the following result.

Corollary 3.1. *Suppose Z is a strictly feasible σ-absolute-optimal solution to v_δ. Then Procedure* INNER-PACKCOVER *in Fig. 3.1 computes an $\alpha\sigma$-absolute-optimal solution $\hat{Z}$ to v_δ in $\mathcal{O}(\frac{1}{\alpha g_\delta(Z)} \cdot \tau \cdot ||M|| \cdot n \ln(n) \cdot \sqrt{\ln(m)})$ time, where $g(Z)$ is the feasibility margin of Z and $||M|| = \max_i \{\lambda_{\max}(M_i)\}$.*

Proof. Fix $\gamma = \frac{\sigma}{g_\delta(Z)} \geq \frac{\langle C, Z \rangle - v_\delta}{g(Z)}$. Let $\hat{X}$ denote a $\alpha\sigma$-approximate saddle-point for $v_\delta(\gamma)$. Then Lemma 3.1 implies that

---

INNER-PACKCOVER

> **Inputs:** $\alpha \in (0,1)$, $\delta \in (0,\delta_0)$, $\sigma > 0$, and a σ-absolute-optimal, strictly feasible solution Z
> **Outputs:** $\alpha\sigma$-absolute-optimal feasible solution $\hat{Z}$
> Set $\gamma = \frac{\sigma}{g(Z)}$.
> Call the NESTEROV procedure to find $\bar{X}$, an absolute $\alpha\sigma$-optimal solution to $v_\delta(\gamma)$.
> Set $\hat{X} = \frac{1}{g(Z)+g^-(\bar{X})}(g(Z)\bar{X} + g^-(\bar{X})Z)$.
> **return** $\hat{X}$

Fig. 3.1 The inner iteration of the packing–covering SDP algorithm

$$\bar{X} = \frac{g(Z)\hat{X} + g^-(\hat{X})Z}{g(Z) + g^-(\hat{X})}$$

is $\alpha\sigma$-optimal feasible solution for v_δ. Since $\frac{\gamma}{\alpha\sigma} = \frac{1}{\alpha g_\delta(Z)}$, Theorem 3.1 implies that $\hat{X}$ (and, therefore, $\bar{X}$) can be computed using Nesterov saddle-point algorithm in $O\left(\frac{1}{\alpha g_\delta(Z)} \cdot \tau \cdot \|M\| \cdot n\ln(n) \cdot \sqrt{\ln(m)}\right)$ time. $\qquad\square$

Let $\varepsilon^{(0)} = \frac{\theta_U - \theta_L}{\theta_L}$ denote the initial relative error. Then setting $\alpha = \frac{\varepsilon}{\varepsilon^{(0)}}$ and $Z = \bar{Y}$ results in $\mathcal{O}\left(\frac{\varepsilon^{(0)}}{\delta\varepsilon} \cdot \tau \cdot \|M\| \cdot n\ln(n) \cdot \sqrt{\ln(m)}\right)$ running time. Thus, we have the following result.

Corollary 3.2. *Procedure* INNER-PACKCOVER *computes an ε-optimal δ-feasible solution for the packing–covering SDP* (3.1) *in* $\mathcal{O}\left(\delta^{-1}\varepsilon^{-1}p(n,m) \cdot \tau \cdot \|M\| \cdot n\ln(n) \cdot \sqrt{\ln(m)}\right)$ *time.*

3.3.2   Bisection Search

Our algorithm PACKCOVER displayed in Fig. 3.2 embeds the INNER-PACKCOVER in a bisection search scheme. We first check if the relative error of $\bar{Y}$ is less than the target relative error ε. (Note that $\bar{Y}$ is feasible for v_δ for all $\delta > 0$.) If not, we enter a loop that successively improves the relative error.

Theorem 3.2. PACKCOVER *computes a ε-optimal δ-feasible solution for the packing–covering SDP* (3.1) *in* $\mathcal{O}\left(\delta^{-1}\varepsilon^{-1-\kappa} \cdot \tau \cdot \|M\| \cdot n\ln(n) \cdot \sqrt{\ln(m)}\right)$ *time.*

Proof. Our first claim is θ_L at the end of each iteration k is lower bound for v_δ. This is clearly the case for $k = 0$. Suppose this hypothesis is true for the index k. Then from Corollary 3.1 it follows that $\hat{Z}$ is $\alpha\sigma^{(k)}$-absolute optimal solution for v_δ. Thus, $\langle C, \hat{Z}\rangle > \theta_L + \frac{1+\alpha}{2} \cdot \sigma^{(k)}$ implies that

$$\theta^* \geq \langle C, \hat{Z}\rangle - \alpha\sigma^{(k)} \geq \theta_L + \frac{1+\alpha}{2} \cdot \sigma^{(k)} - \alpha\sigma^{(k)} = \theta_L + \frac{1-\alpha}{2} \cdot \sigma^{(k)}.$$

Thus, $\theta_L + \frac{1-\rho}{2} \cdot (\theta_U - \theta_L)$ is a valid lower bound for θ^*.

PACKCOVER

Inputs: $C, (M_i, b_i), i = 1, \ldots, m, \varepsilon > 0, \theta_L > 0, \overline{Y}$, a strictly feasible solution to v_δ

Outputs: ε-optimal, δ-feasible solution Z for the packing-covering SDP (1).

$\alpha \leftarrow \frac{1}{3}, k \leftarrow 0, Z^{(k)} \leftarrow \overline{Y}, \sigma^{(0)} \leftarrow \langle C, Z^{(0)} \rangle, \varepsilon^{(0)} \leftarrow \sigma^{(0)}/\theta_L$

while $(\sigma^{(k)} > \varepsilon\theta_L)$

    do

        $\hat{Z} \leftarrow \text{INNER-PACKCOVER}(\alpha, \delta, \sigma^{(k)}, Z^{(k)})$

        if $\left(\langle C, \hat{Z} \rangle > \theta_L + \frac{1+\alpha}{2} \cdot \sigma^{(k)} \right)$

        then

$$\theta_L \leftarrow \theta_L + \frac{1-\alpha}{2} \cdot \sigma^{(k)}$$

        $Z^{(k+1)} \leftarrow (1 - \beta^k)\hat{Z} + \beta^k Z^{(0)}$

        $\sigma^{(k+1)} \leftarrow \langle C, Z^{(k+1)} \rangle - \theta_L$

        $k \leftarrow k + 1$

    return $Z^{(k)}$

Fig. 3.2 The packing–covering SDP routine. The problem data C, M_i, b_i and fixed feasibility parameter δ are assumed to be global

Note that in Procedure PACKCOVER $\sigma^{(k)}$ denotes the estimate of absolute error of $Z^{(k)}$. Let $\theta_L^{(k)}$ denote the lower bound at the end of iteration k and define the relative error estimate $\varepsilon^{(k)} = \sigma^{(k)}/\theta_L^{(k)}$. We claim that the sequence of relative error estimates $\{\varepsilon^{(k)}\}$ satisfy the recursion

$$\varepsilon^{(k+1)} \leq \frac{1+\alpha}{2}(1 - \beta^{k+1})\varepsilon^{(k)} + \beta^{k+1}\varepsilon^{(0)}. \tag{3.3}$$

Assume that this relation holds for k. Let $\theta_L^{(k+1)}$ denote the lower bound at the end of iteration $k+1$. First consider the case where $\langle C, \hat{Z} \rangle > \theta_L + \frac{1+\alpha}{2} \cdot \sigma^{(k)}$. In this case,

$$\varepsilon^{(k+1)} = \left(\langle C, Z^{(k+1)} \rangle - \theta_L^{(k+1)} \right)/\theta_L^{(k+1)},$$

$$\leq \left((1 - \beta^k)\left(\langle C, \hat{Z} \rangle - \theta_L^{(k)} - \frac{1+\alpha}{2} \cdot \sigma^{(k)} \right) + \beta^k \langle C, Z^{(0)} \rangle - \theta_L^{(k)} \right)/\theta_L^{(k)},$$

$$\leq \frac{1+\alpha}{2}(1 - \beta^{k+1})\varepsilon^{(k)} + \beta^{k+1}\varepsilon^{(0)},$$

where the first inequality follows from the fact that $\theta_L^{(k+1)} \geq \theta_L^{(k)}$, and the second inequality follows from the fact that $\langle C, \hat{Z} \rangle \leq \langle C, Z^{(k)} \rangle$ and $\theta_L^{(k)} \geq \theta_L^{(0)}$. In the other case,

$$\varepsilon^{(k+1)} = \left(\langle C, Z^{(k+1)} \rangle - \theta_L^{(k+1)} \right) / \theta_L^{(k+1)},$$

$$\leq \left((1 - \beta^k)(\langle C, \hat{Z} \rangle - \theta_L^{(k)}) + \beta^k \langle C, Z^{(0)} \rangle - \theta_L^{(k)} \right) / \theta_L^{(k)},$$

$$\leq \frac{1+\alpha}{2} (1 - \beta^{k+1}) \varepsilon^{(k)} + \beta^{k+1} \varepsilon^{(0)},$$

where the first inequality follows from the fact that $\theta_L^{(k+1)} \geq \theta_L^{(k)}$, and the second inequality follows from the fact that $\langle C, \hat{Z} \rangle \leq \theta_L^{(k)} + \frac{1+\alpha}{2} \cdot \sigma^{(k)}$ and $\theta_L^{(k)} \geq \theta_L^{(0)}$.

We claim that the recursion (3.3) implies that for all $\kappa > 0$, there exists k_0 such that $\varepsilon^{(k)} \leq (\beta^{\frac{1}{1+\kappa}})^k \varepsilon^{(0)}$ for all $k \geq k_0$. For $\beta = \frac{2}{3}$, and $\kappa = \frac{1}{9}$, $k_0 = 118$. Thus, asymptotically, the relative error decays geometrically with a decay factor of $\beta^{1/(1+\kappa)}$; consequently, the **while** loop in Procedure PACKCOVER terminates in at most

$$N = \left\lceil \frac{\ln(\frac{\varepsilon^{(0)}}{\varepsilon})}{\frac{1}{1+\kappa} \ln(\beta)} \right\rceil \leq \left\lceil \frac{\ln(\frac{p(n,m)}{\varepsilon})}{\frac{1}{1+\kappa} \ln(\beta)} \right\rceil$$

iteration. The final step is to bound the overall running time. Note that $g_\delta(Z^{(k)}) \geq \beta^k g_\delta(Z^{(0)}) = \beta^k \delta$. Thus, Corollary 3.1 implies that the overall running time of Procedure PACKCOVER is at most

$$\mathcal{O}\left(\frac{1}{\delta} \cdot \left(\sum_{t=1}^{N} \beta^{-t} \right) \cdot \tau \cdot ||M|| \cdot n \ln(n) \cdot \sqrt{\ln(m)} \right) = \mathcal{O}\left(\frac{1}{\delta} \cdot \beta^{-N} \cdot \tau \cdot ||M|| \cdot n \ln(n) \cdot \sqrt{\ln(m)} \right),$$

where the second expression follows from the fact that $\beta < 1$, and therefore the last term in the sum dominates the sum. The running time estimate follows from the fact that

$$\beta^{-N} = \left(\beta^{\frac{1}{1+\kappa}} \right)^{(1+\kappa)N} = \varepsilon^{-(1+\kappa)}.$$

$\square$

3.4 Applications

Here we describe some SDPs that can be modeled as packing–covering SDPs. For this section, we define **1** as a vector of ones (of the appropriate dimension) and e_i as the ith column of the identity matrix (also of appropriate dimension). In Sect. 3.4.1 we describe the Vertex Cover packing–covering SDP relaxation, in Sect. 3.4.2 we describe the MIN UNCUTpacking–covering SDP relaxation, and in Sect. 3.4.3 we describe a SDP relaxation for the scheduling problem of minimizing weighted completion on two unrelated machines. In this extended abstract, we focus on combinatorial problems, but our methods can also be applied to problems such as CDMA [19] and distance localization [5].

3.4.1 Vertex Cover

In Weighted Vertex Cover problem, we are given an undirected graph $G = (\mathcal{V}, \mathcal{E})$ with $n = |\mathcal{V}|$, $m = |\mathcal{E}|$, and vertex weights $w_i, i \in \mathcal{V}$. A vertex cover is a vertex set $C \subset \mathcal{V}$ such that every edge $(i, j) \in \mathcal{E}$ has $i \in C$ or $j \in C$. For a given cover, C, we define $w(C) = \sum_{i \in C} w_i$. The Weighted Vertex Cover problem is to find the vertex cover with minimum weight. The Vertex Cover SDP relaxation of Kleinberg and Goemans [12] can be described as a packing–covering SDP. The SDP relaxation is of the binary quadratic program,

$$vc(G) = \min \left\{ \frac{1}{2} \sum_{i \in \mathcal{V}} w_i (1 + x_0 x_i) : (x_0 - x_i)(x_0 - x_j) = 0, (i, j) \in \mathcal{E}, x_i \in \{-1, 1\} \right\}.$$

In their SDP, they relax the binary variables x_i to vectors v_i for each i. Then, the SDP and its corresponding packing–covering SDP are as follows.

$$
\begin{aligned}
&\min && \tfrac{1}{2} \sum_{i \in \mathcal{V}} w_i (1 + v_0 v_i) && sd(G) = \min \tfrac{1}{2} \langle C, X \rangle \\
&\text{s.t.} && v_0^\top v_0 = 1 && \text{s.t. } \mathbf{diag}(X) = \mathbf{1} \\
&\forall i && v_i^\top v_i = 1 && \langle F_{ij}, X \rangle = 1 \quad (i, j) \in \mathcal{E} \\
&\text{(a) } (i, j) \in \mathcal{E} && (v_0 - v_i)^\top (v_0 - v_j) = 0 && \langle F_{ij}, X \rangle \geq 1 \quad i, j \in \mathcal{V} \\
&\text{(b) } i, j \in \mathcal{V} && (v_0 - v_i)^\top (v_0 - v_j) \geq 0, && X \succeq 0, \mathbf{Tr}(X) \leq n+1
\end{aligned}
$$

Note that the dimension of the matrices in the packing–covering formulation is $n + 1$. Here, $F_{ij} = \frac{1}{2}(e_i + e_j - e_0)(e_i + e_j - e_0)^\top$ is a rank-one matrix that represents constraints (a) and (b) while using the diagonal restriction. Also, $C = (C_{ij})$ has zeros everywhere except entries $C_{ii} = w_i$ for $i \in \mathcal{V}$ and $C_{0i} = C_{i0} = w_i/2$ for $i \in \mathcal{V}$. The only new constraint is that the trace of the matrix is bound by $n + 1$, which is an implied constraint from the diagonal equality constraints. Using standard SDP solvers is intractable for the strengthened constraints, which results in a running time of $\tilde{O}(n^{6.5})$. We obtain the following runbound.

Corollary 3.3. *An ε-approximate, δ-feasible solution to the packing–covering SDP for Vertex Cover can be found in* $O(\varepsilon^{-1} \cdot \delta^{-1} n^4 \sqrt{\log(n)} + \ell(n, m))$ *time.*

Proof. We can set $\mathcal{X} = \{X \succeq 0 : \mathbf{Tr}(X) \leq n+1\}$ and choose $\overline{Y}$ to be the matrix corresponding to the LP-based two-approximation solution. Then, since the SDP $SD(G)$ is an upperbound on the LP relaxation[1] (Prop. 1, [12]), we have $\theta_U = \langle C, \overline{Y} \rangle$ and can set $\theta_L = \theta_U/2$. Thus, $p(n, m) = 2$, $\|M\| = O(1)$, and the number of constraints is $O(n^2)$. $\qquad \square$

---

[1] To obtain a solution to the usual LP from the SDP, set $y_i = \frac{v_0^\top v_i + 1}{2}$.

We use the more conservative $n\ln(n) = \mathcal{O}(n^3)$, although we note that in light of [6,9], only a few eigenvalues seem to be required in practice, so a bound of $n\ln(n) = \mathcal{O}(n^2)$ is probably correct.

3.4.2 MIN UNCUT

The MIN UNCUT problem is a version of maximum satisfiability where the clauses are either–or or negated either–or. One formulation is to use $b_i, i = 1, \ldots, n$ as boolean variables so that the m clauses are either of form $b_i \oplus b_j = 1$ or $b_i \oplus b_j = 0$. Agarwal, et al. [1] formed a vector relaxation to solve the associated graph problem, and used the rounding algorithm of Arora et al. [3] to obtain a $\mathcal{O}(\sqrt{n})$-approximation algorithm. In the graph formulation used in [1], there is a node for each clause, b_i, and another node for its negation, b_{-i}, i.e., $V = \{1, \ldots, n, -1, \ldots, -n\}$. Each edge in the graph represents the two pairs of boolean variables that must have the same value in order to satisfy each either–or constraint. Thus, $\mathcal{E} = \mathcal{E}_0 \cup \mathcal{E}_1$ where $\mathcal{E}_0 = \{(i, j), (-i, -j) : b_i \oplus b_j = 0\}$ and $\mathcal{E}_1 = \{(-i, j), (i, -j) : b_i \oplus b_j = 1\}$. Thus, $|V| = 2n$ and $|\mathcal{E}| = 2m$. A *symmetric cut* in the graph $\mathcal{G} = (V, \mathcal{E})$ is a node set $S \subset V$ so that $V - S = \{-i : i \in S\}$ and corresponds to a legitimate setting of boolean variables with all the nodes in S being set to 1 and the ones in $V - S$ set to 0. Note that if the edge (i, j) crosses the symmetric cut, then so does the edge $(-i, -j)$. Thus, the edges crossing a symmetric cut represent twice the number of constraints unsatisfied by this variable setting. So finding such a minimum symmetric cut solves the MIN UNCUT problem (and is the origin of the name). To form a vector relaxation, vectors v_i are assigned to each variable, b_i, and $-v_i$ to b_{-i}. Then there is an objective term for each $(i, j) \in \mathcal{E}, \|\mathbf{sgn}(i)v_i - \mathbf{sgn}(j)v_j\|^2$. To use the rounding algorithm of [3], the triangle inequalities must be satisfied, which with the substitution of $-v_i$ for v_{-i} results in the eight possible sign settings in the equality $\| \pm v_i \pm v_j \|^2 + \| \pm v_j \pm v_k \|^2 \geq \| \pm v_i \pm v_k \|^2$ for all i, j, k.

   To model the vector formulation as a packing–covering SDP the Gram matrix $X = V^\top V$ is used, where the columns of V are the vectors v_i. As is well known, a matrix is Gram if and only if it is positive semidefinite. The objective matrix is $C = \frac{1}{4} \sum_{(i,j) \in \mathcal{E}} (\mathbf{sgn}(i)e_i - \mathbf{sgn}(j)e_j)(\mathbf{sgn}(i)e_i - \mathbf{sgn}(j)e_j)^\top$. The constraint matrices for the triangle inequalities are $T_{ijk} := aa^\top + bb^\top - cc^\top$ where $a = (\mathbf{sgn}(i)e_{|i|} - \mathbf{sgn}(j)e_{|j|})$, $b = (\mathbf{sgn}(j)e_{|j|} - \mathbf{sgn}(k)e_{|k|})$, $c = (\mathbf{sgn}(i)e_{|i|} - \mathbf{sgn}(k)e_{|k|})$. Putting this all together results in the following equivalent formulations.

$$\begin{aligned} &\min \tfrac{1}{4} \sum_{(i,j) \in E} c_{ij} \|v_i - v_j\|^2 \\ &\text{s.t. } \|v_i\|^2 = 1, \forall i \\ &\forall i, j, k, \ \|v_i - v_j\|^2 + \|v_j - v_k\|^2 \\ &\qquad \geq \|v_i - v_k\|^2, \end{aligned} \qquad \begin{aligned} &UC(\mathcal{G}) = \min \tfrac{1}{4} \langle C, X \rangle \\ &\text{s.t. } x_{ii} = 1, i \in V \\ &\tfrac{1}{2} \langle T_{ijk}, X \rangle \geq 1, \pm i, \pm j, \pm k \in V \\ &X \succeq 0, \mathbf{Tr}(X) = n. \end{aligned}$$

We then have the following corollary.

Corollary 3.4. *A ε-optimal, δ-feasible solution to the* MIN UNCUT*SDP relaxation can be solved in* $\tilde{O}(\varepsilon^{-1} \cdot \delta^{-1} \cdot n^4 + \ell(n,m))$

Proof. As an initial solution, we can use the solution from the algorithm of Garg et al. [7], which generates a $\log(n)$-approximate solution, $\bar{Y}$. Their algorithm involves an LP solve followed by (essentially) a shortest path computation. Then, we can set $\theta_U = \langle C, \bar{Y} \rangle$ and $\theta_L = \theta_U / \log(n)$. Note that $\tau = n$, $\|M\| = \mathcal{O}(1)$ and $n \ln(n) = \tilde{O}(n^3)$. The overall time bound follows from Theorem 3.2. $\qquad\square$

3.4.3 Scheduling

Skutella [18] described a semidefinite programming-based algorithm to solve the scheduling problem of minimizing weighted completion time on two unrelated machines. To use this formulation, we let $\mathcal{J}$ to denote the set of jobs to be completed, and each job, $j \in \mathcal{J}$ has a process time on the first machine of p_{-1j} and p_{0j} on the second machine. For convenience, for $i \in \{-1,0\}$ and for each $j \in \mathcal{J}$, we define $J(i,j) = \{k \in \mathcal{J} : \dfrac{w_j}{p_{ij}} > \dfrac{w_k}{p_{ik}}$ or $\dfrac{w_j}{p_{ij}} = \dfrac{w_k}{p_{ik}}, j < k\}$. The SDP used can be straightforwardly be written as the following packing–covering SDP. The dimension of all matrices are $n+2$ and the indices are integers that range from -1 to n, where the indices -1 and 0 represent the two machines and the indices 1 through n represent the jobs to be scheduled.

To round the SDP, a solution, v^*, from the convex quadratic program in [18] is required. For each $i, j \in \mathcal{J}$, we let $a_{ij} = \frac{1 + v_i^* v_j^*}{2}$ and

$$\sigma = \sum_j w_j \Big(\sum_{i=-1,0} a_{ij} \cdot \Big(\frac{1 + a_{ij}}{2} p_{ij} + \sum_{k \in J(j,i)} a_{ik} \cdot p_{ik}\Big)\Big).$$

Let C is the matrix so that for all symmetric matrices X,

$$\langle C, X \rangle = \sum_j w_j \Big(\sum_{i=-1,0} \Big(\frac{X_{jj} + X_{ij}}{2} p_{ij} + \frac{1}{4} \sum_{k \in J(j,i)} (X_{jk} + X_{ij} + X_{ik} + X_{kk}) p_{ik}\Big)\Big).$$

For $i, j \in \mathcal{J}$ and $k \in \{-1,0\}$ define $B_{ijk} = 1/2(-1)^k (e_j e_k^\top + e_k e_j)^\top$. The packing–covering SDP is then as follows.

$$\begin{aligned}
\min \ & \langle C, X \rangle \\
\text{s.t. } & \mathbf{diag}(X) = \mathbf{1} \\
& \langle A_{ijk}, X \rangle \geq 1, i, j \in \mathcal{J}, k \in \{-1,0\} \\
& \langle \tfrac{1}{\sigma} C, X \rangle \geq 1 \\
& \langle B_{ijk}, X \rangle = 1, i, j \in \mathcal{J}, k \in \{-1,0\} \\
& X \succeq \mathbf{0}, \mathbf{Tr}(X) \leq n.
\end{aligned} \qquad (3.4)$$

Here the matrices $A_{ijk} = (e_i + e_j + e_k)(e_i + e_j + e_k)^\top$ and thus we have $\|M\| = \mathcal{O}(1)$. A feasible solution can be found via the constant factor LP-based approximation algorithm of Schulz and Skutella to find an initial solution as well as a upper and lower bound. Also, $n\ln(n) = \mathcal{O}(n^3)$, so we have the following corollary from Theorem 3.2.

Corollary 3.5. *An ε-approximate, δ-feasible solution to* (3.4) *can be found in* $\tilde{\mathcal{O}}(\varepsilon^{-1} \cdot \delta^{-1} \cdot n^4)$ *time.*

Acknowledgements G. Iyengar was supported in part by NSF grants CCR-00-09972, DMS-01-04282, and ONR grant N000140310514. D.J. Phillips was supported in part by NSF grant DMS-0703532 and a NASA/VSGC New Investigator grant. C. Stein was supported in part by NSF grants CCF-0728733 and CCF-0915681.

References

1. Agarwal, A., Charikar, M., Makarychev, K., Makarychev, Y.: $O(\sqrt{\log n})$ approximation algorithms for Min UnCut, Min 2CNF Deletion, and directed cut problems. In: Proceedings of the Thirty-Seventh Annual ACM Symposium on Theory of Computing, pp. 573–581. ACM, New York (2005)
2. Arora, S., Kale, S.: A combinatorial, primal-dual approach to semidefinite programs. In: Proceedings of the 39th Annual ACM Symposium on Theory of Computing, pp. 227–236. ACM, New York (2007)
3. Arora, S., Rao, S., Vazirani, U.: Expander flows, geometric embeddings, and graph partitionings. In: Proceedings of the 36th Annual ACM Symposium on Theory of Computing, pp. 222–231. ACM, New York (2004)
4. Bienstock, D.: Potential function methods for approximately solving linear programming problems: theory and practice. In: International Series in Operations Research & Management Science, vol. 53. Boston, MA (2002)
5. Biswas, P., Liang, T.-C., Toh, K.-C., Ye, Y., Wang, T.-C.: Semidefinite programming approaches for sensor network localization with noisy distance measurements. IEEE Trans. Autom. Sci. Eng. **3**, 360–371 (2006)
6. d'Aspremont, A.: Smooth optimization with approximate gradient. SIAM J. Optim. **19**, 1171–1183 (2008)
7. Garg, N., Vazirani, V., Yannakakis, M.: Approximate max-lfow min-(multi)cut theorems and their applications, SIAM Journal on Computing, **25**, 235–251 (1996)
8. Goldberg, A.V., Oldham, J.D., Plotkin, S.A., Stein, C.: An implementation of a combinatorial approximation algorithm for minimum-cost multicommodity flow. In: Proceedings of the 4th Conference on Integer Programming and Combinatorial Optimization, pp. 338–352 (1998). Published as Lecture Notes in Computer Science 1412, Springer-Verlag
9. Iyengar, G. Phillips, D.J., Stein, C. Approximating semidefinite packing programs, SIAM Journal on Optimization, **21**, 231–268 (2011)
10. Iyengar, G., Phillips, D.J., Stein, C.: Feasible and accurate algorithms for covering semidefinite programs. In: Proceedings of the 12th Scandinavian Workshop on Algorithms and Theory, 2010
11. Klein, P., Lu, H.-I.: Efficient approximation algorithms for semidefinite programs arising from MAX CUT and COLORING. In: Proceedings of the Twenty-eighth Annual ACM Symposium on the Theory of Computing (Philadelphia, PA, 1996), pp. 338–347. ACM, New York (1996)
12. Kleinberg, J., Goemans, M.: The Lovász theta function and a semidefinite programming relaxation of vertex cover. SIAM J. Disc. Math. **11**, 196–204 (1998)

13. Lu, Z., Monteiro, R., Yuan, M.: Convex optimization methods for dimension reduction and coefficient estimation in multivariate linear regression. Math. Program. **131**, 163–194 (2012)
14. Nesterov, Y.: Smooth minimization of nonsmooth functions. Math. Program. **103**, 127–152 (2005)
15. Nesterov, Y.: Smoothing technique and its applications in semidefinite optimization. Math. Program. **110**, 245–259 (2007)
16. Nesterov, Y., Nemirovski, A.: Interior-point polynomial algorithms in convex programming. In: SIAM Studies in Applied Mathematics, Society for Industrial and Applied Mathematics (SIAM), vol. 13. Philadelphia (1994)
17. Plotkin, S., Shmoys, D.B., Tardos, E.: Fast approximation algorithms for fractional packing and covering problems. Math. Oper. Res. **20**, 257–301 (1995)
18. Skutella, M.: Convex quadratic and semidefinite programming relaxations in scheduling. J. ACM **48**, 206–242 (2001)
19. Tan, P., Rasmussen, L.: The application of semidefinite programming for detection in CDMA. IEEE J. Sel. Areas Commun. **19**, 1442–1449 (2001)

Chapter 4
On the Tendency Toward Convexity of the Vector Sum of Sets

Roger E. Howe

There are several instances (e.g., [2, p. 255], [3, p. 255]) in the general equilibrium theory of economics where one is interested in the following question:

Given sets $\{X_i\}_{i=1}^n$ in a vector space V, how close is the vector sum

$$\sum_{i=1}^n X_i = \left\{ \sum_{i=1}^n x_i : x_i \in X_i \right\} \tag{4.1}$$

to being convex?

The standard result on this topic is the Shapley–Folkman theorem [1, 2]. In this note we offer some observations regarding the question, beginning with a proof of the Shapley–Folkman result and proceeding to various refinements.

As above, let $\{X_n\}_{i=1}^n$ be bounded sets in a vector space V of dimension ℓ. We will assume that the X_i are closed (and hence compact). (This is no loss of generality for Shapley–Folkman and is convenient for our arguments.) Also for convenience and without loss of generality in what follows, we may assume that $0 \in X_i$ for all i. This is because if $x_i^o \in X_i$, then $0 \in X_i - x_i^o$ and

$$\sum_{i=1}^n (X_i - x_i^o) = \left(\sum_{i=1}^n X_i \right) - \left(\sum_{i=1}^n x_i^o \right). \tag{4.2}$$

So requiring $0 \in X_i$ merely amounts to a translation of the whole situation.

Let $co(X_i)$ denote the convex hull of X_i. It is well known that $co(X_i)$ is closed [4]. Further, the extreme points of $co(X_i)$ all belong to X_i. It is also standard that

$$co\left(\sum_{i=1}^n X_i \right) = \sum_{i=1}^n co(X_i). \tag{4.3}$$

R.E. Howe (✉)
Department of Mathematics, Yale University, New Haven, CT, USA
e-mail: howe@math.yale.edu

T. Terlaky and F.E. Curtis (eds.), *Modeling and Optimization: Theory and Applications*, Springer Proceedings in Mathematics & Statistics 21, DOI 10.1007/978-1-4614-3924-0_4, © Springer Science+Business Media, LLC 2012

Thus any point z in $co(\Sigma_{i=1}^n X_i)$ can be written in the form

$$z = \sum_{i=1}^n x_i \quad x_i \in co(X_i). \tag{4.4}$$

Proposition 4.1 (Shapley–Folkman). *In the representation* (4.4) *all but* $\ell = \dim V$ *of the* x_i *may be taken to belong to* X_i *(instead of only* $co(X_i)$*).*

Proof. The proof proceeds by induction on n and ℓ. Put

$$Z^k = co\left(\sum_{i=1}^k X_i\right).$$

for $k \leq n$. A point $z \in Z^n$ can be written

$$z = y + x_n \quad y \in Z^{n-1}, \quad x_n \in co(X_n).$$

Since $0 \in X_n$, also $y \in Z^n$. Hence the whole line segment

$$y + tx_n \quad 0 \leq t \leq 1$$

belongs to Z^n. There are two possibilities: either (a) $z \in Z^{n-1}$, or (b) $z \notin Z^{n-1}$. In the first case, the result for the point $z \in Z^n$ follows from the result for the point $z \in Z^{n-1}$. In the second case, there will be a largest t for which $y + tx_n$ belongs to Z^{n-1}. For this t, $y + tx_n = y'$ will be on the boundary ∂Z^{n-1} of Z^{n-1}. We may thus write

$$z = y' + x'_n \quad y' \in \partial Z^{n-1}, x' \in co(X_n).$$

We come now to the crucial point. We may choose a supporting hyperplane to Z^{n-1} at y'. Thus, let λ be a linear functional on V such that

$$\max\{\lambda(u) : u \in Z^{n-1}\} = \lambda(y').$$

Let

$$H(\lambda,s) = \{v \in V : \lambda(v) = s\}$$

be the set of parallel hyperplanes defined by λ. Set

$$s_i = \max\{\lambda(x_i) : x_i \in X_i\}.$$

Then by choice of λ, we see that $\lambda(y') = \Sigma_{i=1}^{n-1} s_i$, and

$$H(\lambda,s_i) \cap co(X_i) = co(H(\lambda,s_i) \cap X_i)$$

and

$$Z^{n-1} \cap H(\lambda, \lambda(y')) = \sum_{i=1}^{n-1} H(\lambda, s_i) \cap co(X_i).$$

Up to translation, we see that we are dealing with the situation of the theorem, but in dimension only $\ell - 1$. Thus the result may be assumed for y' with $\ell - 1$ in place of ℓ; and then $z = y' + x_n'$ is the desired representation for z. □

We note a couple of corollaries. First, suppose that the only defect from convexity in the x_i is that they are hollow; precisely, assume that $\partial co(X_i) \subseteq X_i$. Then inductively one sees that $\partial Z^n \subseteq \Sigma_{i=1}^n X_i$. Hence, from proof we conclude

Corollary 4.1. *If $\partial co(X_i) \subseteq X_i$ for all i, then an arbitrary $z \in Z^n = co(\Sigma_{i-1}^n X_i)$ can be written as*

$$z = y + x$$

where $y \in \sum_{i=1}^n X_i$ and $x \in co(X_j)$ for some j. □

Corollary 4.1 obviously involves rather a special situation. However, the same basic aspect of Proposition 4.1 that it exploits can also be used in other ways. For example, given a linear functional λ on V, let $e(\lambda)$ be the number of the sets X_i such that λ assumes its maximum on X_i at more than one point of X_i. For a given X_i, it can be shown that the typical λ (i.e., for λ in a set of full measure in V^*, the dual of V) assumes its maximum only once on X_i. Thus if the X_i are imagined to be chosen "at random," it is plausible that $e(\lambda)$ might be ≤ 1 for all λ. In any case, let us set

$$\tilde{e}(\{X_i^n\}_{i=1}) = \{\max e(\lambda) : \lambda \in V^*, \lambda \neq 0\}.$$

Then a repetition with slight changes of the proof of Proposition 4.1 above gives

Corollary 4.2. *In the expansion (4.4), all but $\tilde{e}(\{X_i\}_{i=1}^n) + 1$ of the x_i's may be chosen to belong to X_i, rather than $co(X_i)$.* □

To further refine Proposition 4.1 we introduce quantitative considerations. Suppose $\|\ \ \|$ is a norm on V. Define the *diameter* $\delta(X_i)$ of the sets X_i by

$$\delta(X_i) = \sup\{\|x - x'\| : x, x' \in X_i\}. \tag{4.5}$$

Thus X_i is contained in the intersection of the balls of radius $\delta(X_i)$ centered at the points of X_i. It follows that

$$\delta(co(X_i)) = \delta(X_i). \tag{4.6}$$

If $v \in V$, and $X \subseteq V$ is a set, define the *distance* $d(v, X)$ from v to X by

$$d(v, X) = \inf\{\|v - x\| : x \in X\}. \tag{4.7}$$

In terms of these definitions we immediately have the following "metrical" version of Proposition 4.1.

Corollary 4.3. *For any $z \in co(\Sigma_{i=1}^{n} X_i)$, we have*

$$d\left(z, \sum_{i=1}^{n} X_i\right) \leq (\dim V)\left(\max_{i} \delta(X_i)\right).$$ □

To see that this estimate is of the correct order of magnitude, let $V = \mathbf{R}^{\ell}$ and let each set X_i consist of the standard basis vectors together with 0. Then

$$co(X_i) = \left\{ \mathbf{v} \in \mathbf{R}^n : \mathbf{v} = \begin{bmatrix} v_1 \\ v_2 \\ \vdots \\ v_n \end{bmatrix} ; 0 \leq v_i, \ and \ \sum_{i=1}^{n} v_i \leq 1 \right\}.$$

Further $Z^n = \Sigma_{i=1}^{n} co(X_i) = n \, co(X_1)$.
Let $\| \quad \|$ be the 1-norm:

$$\|(x_1, \ldots, x_{\ell})\|_1 = \sum_{j=1}^{n} |x_j|.$$

Then for large n (in fact, already for $n \geq \frac{\ell}{2}$), the vector

$$v = \frac{1}{2}(1, 1, 1, \ldots, 1) \qquad (\textit{with } \ell \ 1s)$$

will be in Z^n, but the points of $\Sigma_{i=1}^{n} X_i$ all have integer coordinates. Thus for any x in $\Sigma_{i=1}^{n} X_i$, all the coordinates of $v - x$ are at least $\frac{1}{2}$ in absolute value, whence $d(v, \Sigma_{i=1}^{n} X_i) \geq \frac{\ell}{2}$. On the other hand, $\delta(X_i) = 2$, so $d(v, \Sigma_{i=1}^{n} X_i) \geq \frac{\ell(\max_i \delta(X_i))}{4}$. To get better results, we must make some further assumptions on the X_i. One obvious possibility is that some of the X_i have nonempty interiors. If this holds, then $\Sigma_{i=1}^{n} X_i$ will tend to fill up the inside of its convex hull.

For $x \in \mathbf{R}^{\ell}$ and a number $r > 0$, let $\bar{B}_r(x)$ denote the closed ball of radius r centered at the point x. If X_i has nonempty interior, then it will contain balls of positive radius around some of its points. Let $v(X_i)$ be the radius of the largest closed ball contained in X_i. Then we have

$$v\left(\sum_{i=1}^{n} X_i\right) \geq \sum_{i=1}^{n} v(X_i).$$ (4.8)

This is because,

$$\sum_{i=1}^{n} \bar{B}_{r_i}(x_i) = \bar{B}_R(X),$$

where $R = \Sigma r_i$ and $X = \Sigma_{i=1}^{n} x_i$. This fact lies at the base of our second result, which is also quite simple.

Proposition 4.2. *Set*

$$\max_i \delta(X_i) = \delta_o.$$

Suppose we can find certain of the X_i, say $X_1, X_2, \ldots, X_m$, with $m \leq n$ such that $\Sigma_{i=1}^m v(X_i) \geq (\dim V)\delta_o$. Then $\Sigma_{i=1}^n X_i$ contains all points sufficiently far inside $Z^n = co(\Sigma_{i=1}^n X_i)$. Precisely, suppose $z \in Z^n$ and

$$d(z, \partial Z^n) \geq m\delta_o.$$

Then $z \in \Sigma_{i=1}^n X_i$.

Proof. We may write

$$\sum_{i=1}^n X_i = \left(\sum_{i=1}^m X_i \right) + \left(\sum_{i=m+1}^n X_i \right) = Y_1 + Y_2.$$

By our assumptions and (4.8), we have $v(Y_1) \geq \ell \delta_o$. Thus Y_1 contains a ball of radius $\ell \delta_o$. By a translation of Y_1, an operation which will not affect our argument, we may assume that Y_1 contains the ball of radius $\ell \delta_o$ around the origin. Then also $\delta(Y_1) \leq m\delta_o$, so Y_1 is contained in the ball of radius $m\delta_o$ around the origin.

By Corollary 4.3, no point of $co(Y_2)$ is at distance greater than $\ell \delta_o$ from Y_2. It follows that $co(Y_2) \subseteq Y_1 + Y_2$. Further,

$$Z^n = co(Y_1 + Y_2) = co(Y_2) + co(Y_1) \subseteq coY_2 + \bar{B}_{m\delta_o}(0)$$

It is intuitively clear from this that any point of Z^n at distance more than $m\delta_o$ from ∂Z^n must be in $co(Y_2)$, hence in $Y_1 + Y_2$. Here is the argument. Suppose $z_o \notin co(Y_2)$. Then we can find a linear functional λ in V^*, the dual of V, separating z_o from $co(Y_2)$. We may assume (recalling that $co(Y_2)$ contains the origin)

$$\lambda(z_o) > \max\{\lambda(y) : y \in co(Y_2)\} \geq 0.$$

By the Hahn–Banach Theorem, we can find $v \in B_{m\delta_o}(z)$ such that

$$\lambda(v) = \lambda(z_o) + \|\lambda\|^* m\delta_o > max\{\lambda(z), z \in coY_2 + \bar{B}_{m\delta_o}\} \geq max\{\lambda(z) : z \in Z^n\}$$

where $\|\lambda\|^*$ indicates the norm of λ in the norm on V^* dual to our given norm $\| \ \|$ on V. Thus clearly $v \notin Z^n$. If we draw the line segment from z_o to v, then the end at z_o belongs to Z^n, but the end at v does not. Hence at some place, the segment pierces ∂Z^n. Since the length of the segment is exactly $m\delta_o$ by construction, the distance from z_o to ∂Z^n is less than $m\delta_o$, as desired.   □

Thus, existence of interior points in the X_i causes $\Sigma_{i=1}^n X_i$ to fill up its convex hull. However, even if each X_i is the closure of its interior, there may be points in $co(\Sigma_{i=1}^n X_i)$ which stay a fixed distance away from $\Sigma_{i=1}^n X_i$, as the example in Fig. 4.1 shows. Evidently the problem there is the angularity of X. Thus if we put some sort

Fig. 4.1 Example illustrating points in $co(\sum_{i=1}^{n}X_i)$ which stay a fixed distance away from $\sum_{i=1}^{n}X_i$ when each X_i is the closure of its interior

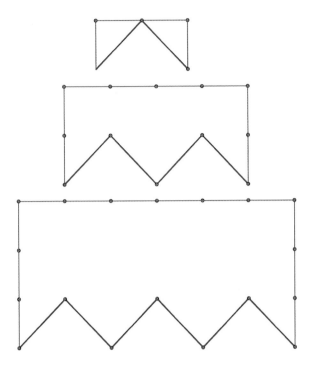

of smoothness condition on the X_i, we might expect to improve on Proposition 4.2. This is the case; here is one formulation of such a result.

We take $V = \mathbf{R}^{\ell}$ with the usual Euclidean norm.

Proposition 4.3. *Suppose X_i is a union of balls of some fixed radius $r_i \geq 0$. Put $s = \sum_{i=1}^{n}r_i$, and put $\delta_o = \max_i\{\delta(X_i) - 2r_i\}$. Assume $s \geq \ell\delta_o$. Then for any point $z \in Z^n = co(\sum_{i=1}^{n}X_i)$ we have*

$$d\left(z, \sum_{i=1}^{n}X_i\right) \leq \sqrt{s^2 + (\ell\delta_o)^2} - s = \frac{(\ell\delta_o)^2}{\sqrt{s^2 + (\ell\delta_o)^2} + s} \leq \frac{1}{2s}(\ell\delta_o)^2. \qquad (4.9)$$

Remarks. (a) If X_i is the interior of a smooth (twice continuously differentiable) hypersurface, then X_i will be the union of balls of some sufficiently small radius.
(b) By (4.9) we see that if a large number of the X_i are relatively rotund, so that s is large, then $\sum_{i=1}^{n}X_i$ fills most of Z^n, in the sense that all points of Z^n are fairly close to it.

Proof. Let $\tilde{X}_i$ be the set of centers of balls of radius r_i contained in X_i. Then

$$X_i = \tilde{X}_i + B_{r_i}(0)$$

and

$$\delta(X_i) = \delta(\tilde{X}_i) + 2r_i.$$

Hence

$$\sum_{i=1}^{n} X_i = \sum_{i=1}^{n} (\tilde{X}_i + B_{r_i}(0)) = \sum_{i=1}^{n} \tilde{X}_i + B_s.$$

Thus, if we set $\tilde{Z}^n = \Sigma_{i=1}^{n} co(\tilde{X}_i)$, we can write

$$Z^n = \tilde{Z}^n + B_s.$$

Since we assume $s \geq \ell \delta_o$, we see by Corollary 4.3 that

$$\tilde{Z}^n \subseteq \sum_{i=1}^{n} X_i.$$

Thus if $z \in Z^n - \Sigma_{i=1}^{n} X_i$, then in particular $z \in Z^n - \tilde{Z}^n$. Let $\tilde{z}$ be the point of $\tilde{Z}^n$ closest to z. Note that $\tilde{z}$ is unique, since $\tilde{Z}^n$ is convex and Euclidean balls are strictly convex. The hyperplane H passing through $\tilde{z}$ and perpendicular to $z - \tilde{z}$ will support $\tilde{Z}^n$.

We are now in a situation like the inductive situation in Proposition 4.1. The hyperplane H may be described as the set of all vectors $\mathbf{v}$ in $\mathbf{R}^\ell$ such that $(z - \tilde{z}) \cdot \mathbf{v} = (z - \tilde{z}) \cdot \tilde{z}$. Here $\mathbf{u} \cdot \mathbf{v}$ denotes the inner product of vectors $\mathbf{u}$ and $\mathbf{v}$ in $\mathbf{R}^\ell$. For each set $\tilde{X}_i$, let t_i be the maximum value of $(z - \tilde{z}) \cdot x$ over all points x in $\tilde{X}_i$, and let $\tilde{Y}_i$ be the subset of $\tilde{X}_i$ where this maximum is achieved. Then by construction of $\tilde{z}$, $\Sigma_{i=1}^{n} t_i = (z - \tilde{z}) \cdot \tilde{z}$, and

$$\tilde{Z}^n \cap H = \sum_{i=1}^{n} co(\tilde{Y}_i).$$

Applying Corollary 4.3 to the $\tilde{Y}_i$, we conclude that there is a point $\tilde{y}$ of $\Sigma_{i=1}^{n} \tilde{Y}_i \subset \Sigma_{i=1}^{n} \tilde{X}_i$ within distance $\ell \delta_o$ (actually, $(\ell - 1)\delta_o$) of $\tilde{z}$. Since $\tilde{y} - \tilde{z}$ is orthogonal to $z - \tilde{z}$, we can estimate

$$||z - \tilde{y}||^2 = ||z - \tilde{z}||^2 + ||\tilde{z} - \tilde{y}||^2 \leq s^2 + (\ell \delta_o)^2.$$

On the other hand, we know that the ball $\bar{B}_s(\tilde{y})$ is contained in $\Sigma_{i=1}^{n} X_i$. This means that z is within $||z - \tilde{y}|| - s \leq \sqrt{s^2 + (\ell \delta_o)^2} - s$ of $\Sigma_{i=1}^{n} X_i$. $\square$

To summarize, we have in Propositions 4.1–4.3 the following progression. First, if you take the vector sum of small sets, no point of the convex hull of the sum is far away from a point of the sum. Second, if an appreciable number of the summands have interior, then the sum tends to completely fill the inside of its convex hull. Third, if an appreciable number of the summands are relatively rotund, then the sum tends to fill almost all of its convex hull. To finish the discussion it seems appropriate to show that in some cases, even if the original X_i do not have interior, summing certain of them may create sets with nonempty interior, so that, after a partitioning of the sum $\Sigma_{i=1}^{n} X_i$ into certain subsums, Proposition 4.2 or 4.3 may be applicable.

First, we remind the reader of the folklore fact that summing sets of positive measure creates sets with interior. Then we show that summing sets which are small, but still positive dimensional (e.g., sets containing continuous arcs) creates sets of positive measure. Finer results of this nature certainly exist, but these seem to exemplify fairly clearly our theme, the tendency to convexity of vector sums.

Suppose $X \subseteq V$ is a set of positive measure $\mu(X)$. A point $x \in X$ is called a *point of density* of X if

$$\lim_{r \to 0} \frac{\mu(X \cap \bar{B}_r(x))}{\mu(\bar{B}_r(x))} = 1. \tag{4.10}$$

Here as before $\bar{B}_r(x)$ is the (closed) ball of radius r around x in the given norm on V. A classical result [6] of Euclidean harmonic analysis guarantees that the set of points of X which are not points of density is of measure zero. This fact immediately implies the following known result.

Proposition 4.4. *If X_1 and X_2 are sets of positive measure, then $X_1 + X_2$ has nonempty interior. More precisely, if $x_i \in X_i$ are points of density for $i = 1, 2$, then $X_1 + X_2$ contains a neighborhood of $x_1 + x_2$.*

Proof. By a translation, we may just as well assume that $x_1 = x_2 = 0$. Then given $\varepsilon > 0$, we can find a radius $r > 0$ such that

$$\mu(X_i \cap \bar{B}_r(0)) \geq (1 - \varepsilon)\mu(\bar{B}_r(0)).$$

Choose $\upsilon \in V$, with $\|\upsilon\| = \alpha r$, with $0 < \alpha < 1$. Let $(\bar{B}_r(\upsilon))'$ denote the complement of $\bar{B}_r(\upsilon)$. Then

$$\mu(X_i \cap \bar{B}_r(0) \cap \bar{B}_r(\upsilon)) \geq \mu(X_i \cap \bar{B}_r(0)) - \mu(\bar{B}_r(0) \cap (\bar{B}_r(\upsilon))') \tag{4.11}$$

$$\geq (1 - \varepsilon)\mu(\bar{B}_r(0)) - (\mu(\bar{B}_r(0)) - \mu(\bar{B}_r(0) \cap \bar{B}_r(\upsilon))) \tag{4.12}$$

$$\geq -\varepsilon\mu(\bar{B}_r(0)) + \mu(\bar{B}_{(1-\alpha)r}(0)) \tag{4.13}$$

$$= ((1 - \alpha)^\ell - \varepsilon)\mu(\bar{B}_r(0)) \tag{4.14}$$

(where $\ell = \dim V$) $\quad \geq ((1 - \alpha)^\ell - \varepsilon)\mu(\bar{B}_r(0) \cap \bar{B}_r(\upsilon)). \tag{4.15}$

We see that if α and ε are small enough, then $(1 - \alpha)^\ell - \varepsilon$ is $> 1/2$. By symmetry, the same reasoning applied to $(X_i + \upsilon) \cap \bar{B}_r(0) \cap \bar{B}_r(\upsilon)$. Hence

$$X_1 \cap (X_2 + \upsilon) \cap \bar{B}_r(0) \cap \bar{B}_r(\upsilon)$$

is nonempty. Pick a point z in it. Then

$$z = x_1 = x_2 + \upsilon$$

or

$$v = z - x_2 = x_1 - x_2.$$

Repeating the above reasoning with $-X_2$ in place of X_2, which one checks is all right, yields $v \in X_1 + X_2$. Since v was arbitrary subject to being sufficiently close to 0, the proposition is proved.   □

Finally, we consider the situation when we have $\ell = \dim V$ continuous arcs inside certain of our sets. For purposes of the proposition, we may as well assume that the sets reduce to just the arcs.

Proposition 4.5. *Let c_i, $0 \le i \le \ell = \dim V$ be ℓ continuous arcs in V. That is,*

$$c_i = [0, 1] \to V$$

are continuous maps. Assume for simplicity that $c_i(0) = 0$. Then the volume of the vector sum of the sets $c_i([0, 1])$, the images of the c_i, is at least as large as the volume of the parallopiped spanned by the points $c_i(1)$. Moreover, assuming that this volume is positive, the sum $\Sigma_{i=1}^{\ell} c_i([0, 1])$ has nonempty interior.

Proof. With no loss of generality we may assume that $V = \mathbf{R}^{\ell}$ and $c_i(1)$ is the i-th standard basis vector. We will in fact show that for any point x in the unit cube $I^{\ell} = [0, 1]^{\ell} \subseteq \mathbf{R}^{\ell}$, there is an integral translate $x + z$ of x, with $z \in \mathbf{Z}^{\ell}$ such that $x + z \in \Sigma_{i=1}^{n} \operatorname{im}(c_i) = Y$. Then taking the intersections of Y with the translates $I^{\ell} + z$ of the unit cube, and translating them by $-z$, we find they completely cover I^{ℓ} and so have altogether measure at least 1, as desired.

Our assertion is equivalent to the statement that the translates $Y + z, z \in \mathbf{Z}^{\ell}$, cover $\mathbf{R}^n$. Define a mapping $C : \mathbf{R}^n \to \mathbf{R}^n$ by

$$C(t + z) = \sum_{i=1}^{n} c_i(t_i) + z \quad t \in I^{\ell}, z \in \mathbf{Z}^{\ell}.$$

It is easy to see that C is well defined and continuous, and our assertion is seen to be equivalent to the statement that C is surjective.

The map C satisfies

$$C(x + z) = C(x) + z, \qquad x \in \mathbf{R}^{\ell}, z \in \mathbf{Z}^{\ell}.$$

In particular C is a mapping of bounded displacement, in the sense that $\|C(x) - x\| \le M$, for some fixed number M. Mappings of bounded displacement of $\mathbf{R}$ to itself are known to be necessarily surjective. This seems to be a folklore result. I was not able to locate it in the literature, so here is a sketch of why it is true.

Let φ be a continuous nonnegative function on $\mathbf{R}$ such that $\varphi(t) = 0$ for $t \le 1$ and $\varphi(t) = 1$ for $t \ge 2$. Pick a very large number r and define a map D_r on $\mathbf{R}^{\ell}$ by

$$D_r(x) = \left(1 - \varphi\left(\frac{\|x\|}{r}\right)\right) c(x) + \varphi\left(\frac{\|x\|}{r}\right) x.$$

Then $D_r(x) = C(x)$ for $\|x\| \leq r$ and $D_r(x) = x$ for $\|x\| \geq 2r$. In particular, for large enough r, D_r defines a map of $\bar{B}_{3r}$ to itself, and this map is the identity on the sphere $\partial\bar{B}_{3r}$. By a well-known variant of the Brouwer fixed point theorem [3], D_r maps $\bar{B}_{3r}$ onto itself. In the limit as $r \to \infty$, we see that C maps $\mathbf{R}^\ell$ onto itself. This shows that $\Sigma_{i=1}^n c_i([0,1])$ must have positive measure.

This argument also implies that $\Sigma_{i=1}^n c_i([0,1])$ has nonempty interior. Indeed, an alternative statement of the result just proven is that the integer translates of the set $\Sigma_{i=1}^n c_i([0,1])$ fills up all of $\mathbf{R}^\ell$. These translates form a countable collection of closed sets. The Baire Category Theorem [5] therefore implies that at least one of the translates of $\Sigma_{i=1}^n c_i([0,1])$ must have nonempty interior. However, a translate of a set in $\mathbf{R}^\ell$ can have interior only if the set itself does. $\qquad\square$

References

1. Arrow, K., Hahn, F.: General Competitive Analysis. Holden Day, San Francisco (1971)
2. Hildenbrand, W., Kirman, A.: Introduction to Equilibrium Analysis. North Holland Press, Amsterdam (1976)
3. Nikaido, H.: Convex Structures and Economic Theory. Academic, New York (1968)
4. Rockafellar, R.T.: Convex Analysis. Princeton Landmarks in Mathematics and Physics. Princeton University Press, Princeton (1996)
5. Royden, H.: Real Analysis. Macmillan, New York (1963)
6. Stein, E.: Singular Integrals and Differentiability Properties of Functions. Princeton University Press, Princeton (1970)

Chapter 5
Multiobjective Optimization via Parametric Optimization: Models, Algorithms, and Applications

Oleksandr Romanko, Alireza Ghaffari-Hadigheh, and Tamás Terlaky

5.1 Introduction

Multicriteria decision making or *multicriteria analysis* is a complex process of finding the best compromise among alternative decisions. A decision maker first describes the problem based on relevant assumptions about the real-world problem. After that, alternative decisions are generated and evaluated. Optimization serves as a tool for solving multicriteria analysis problems when those problems are formulated as *multiobjective optimization* problems. Classical meaning of the word "optimization" refers to single-objective optimization, which is a technique used for searching extremum of a function. This term generally refers to mathematical problems where the goal is to minimize (maximize) an objective function subject to some constraints. Depending on the nature and the form of the objective function and the constraints, continuous optimization problems are classified to linear, quadratic, conic, and general nonlinear optimization problems.

Linear optimization (LO) is a highly successful operations research model. Therefore, it was natural to generalize the LO model to handle more general nonlinear relationships. However, this gives rise to many difficulties such as lack

O. Romanko (✉)
Department of Computing and Software, McMaster University,
1280 Main Street West, Hamilton, ON L8S 4K1, Canada
e-mail: romanko@mcmaster.ca

A. Ghaffari-Hadigheh
Department of Mathematics, Azarbaijan University of Tarbiat Moallem,
35 Kms Tabriz Maragheh Road, Tabriz, East Azarbayjan, Iran
e-mail: hadigheha@azaruniv.edu

T. Terlaky
Department of Industrial and Systems Engineering, Lehigh University,
Harold S. Mohler Laboratory, 200 West Packer Avenue, Bethlehem, PA, USA
e-mail: terlaky@lehigh.edu

T. Terlaky and F.E. Curtis (eds.), *Modeling and Optimization: Theory
and Applications*, Springer Proceedings in Mathematics & Statistics 21,
DOI 10.1007/978-1-4614-3924-0_5, © Springer Science+Business Media, LLC 2012

of strong duality, possible non-convexity and consequently problems with global versus local optimums, lack of efficient algorithms and software, etc.

In the recent decade, a new class of convex optimization models that deals with the problem of minimizing a linear function subject to an affine set intersected with a convex cone has emerged as conic optimization. Although the conic optimization model seems to be restrictive, any convex optimization problem can be cast as a conic optimization model and there are efficient solution algorithms for many classes of conic models such as conic quadratic optimization (CQO) and conic linear optimization (CLO). While it sounds counterintuitive, CQO is a subclass of CLO. Conic optimization has many interesting applications in engineering, image processing, finance, economics, combinatorial optimization, etc.

CLO is the extension of LO to the classes of problems involving more general cones than the nonnegative orthant. As our results heavily rely on duality theory (for review of the topic consult [8]), we present both primal and dual formulations of problems belonging to the CLO class. General form of CLO problem is:

$$
\begin{array}{cc}
\textit{Primal problem} & \textit{Dual problem} \\
\min_{x}\{c^T x \ : \ Ax = b, \ x \in \mathcal{K}\} & \max_{y,s}\{b^T y \ : \ A^T y + s = c, \ s \in \mathcal{K}^*\},
\end{array} \qquad (5.1)
$$

where $\mathcal{K} \in \mathbb{R}^n$ is a closed, convex, pointed, and solid cone, $\mathcal{K}^* = \{s \in \mathbb{R}^n : s^T x >= 0 \ \forall x \in \mathcal{K}\}$ is the dual cone of $\mathcal{K}$; $A \in \mathbb{R}^{m \times n}$, $\text{rank}(A) = m$, $c \in \mathbb{R}^n$, $b \in \mathbb{R}^m$ are fixed data; $x, s \in \mathbb{R}^n$, $y \in \mathbb{R}^m$ are unknown vectors. Often $x \in \mathcal{K}$ is also denoted as $x \geq_{\mathcal{K}} 0$. Moreover, $x \geq_{\mathcal{K}} y$ means $x - y \geq_{\mathcal{K}} 0$.

Examples of pointed convex closed cones include:

• The nonnegative orthant:

$$
\mathbb{R}^n_+ = \mathcal{K}_\ell = \{x \in \mathbb{R}^n \ : \ x \geq 0\},
$$

• The quadratic cone (also know as Lorentz cone, second-order cone, or ice-cream cone):

$$
L^n = \mathcal{K}_q = \{x \in \mathbb{R}^n \ : \ x_1 \geq \|x_{2:n}\|\},
$$

• The semidefinite cone:

$$
S^n_+ = \mathcal{K}_s = \{X \in \mathbb{R}^{n \times n} \ : \ X = X^T, \ X \succeq 0\},
$$

• Any linear transformation and finite direct product of such cones.

Each of the three "standard" cones $\mathcal{K}_\ell$, $\mathcal{K}_q$, and $\mathcal{K}_s$ are closed, convex, and pointed cones with nonempty interior. Moreover, each of these cones are self-dual, which means that the dual cone $\mathcal{K}^*$ is equal to the original cone $\mathcal{K}$. The same holds for any (finite) direct product of such cones.

If x and (y, s) are feasible for (5.1), then the *weak duality* property holds

$$
c^T x - b^T y = x^T s \geq 0. \qquad (5.2)
$$

The *strong duality* property $c^T x = b^T y$ does not always hold for CLO problems. A sufficient condition for strong duality is the primal–dual Slater condition, which requires the existence of a feasible solution pair x and (y,s) for (5.1) such that $x \in$ int $\mathcal{K}$ and $s \in$ int $\mathcal{K}^*$. In this case, the primal–dual optimal set of solutions (x,y,s) is

$$
\begin{aligned}
Ax &= b,\ x \in \mathcal{K}, \\
A^T y + s &= c,\ s \in \mathcal{K}^*, \\
x^T s &= 0.
\end{aligned}
\tag{5.3}
$$

System (5.3) is known as the *optimality conditions* or *Karush–Kuhn–Tucker conditions*.

CQO is the problem of minimizing a linear objective function subject to the intersection of an affine set and the direct product of quadratic cones. CQO is the subclass of CLO and, consequently, CQO problems are expressed in the form of (5.1). More information on CLO and CQO problems, their properties, and duality results can be found in [4]. The CQO problem subclasses described in this section include *linear optimization* (LO), *convex quadratic optimization* (QO), quadratically constrained quadratic optimization (QCQO), and *second-order conic optimization* (SOCO). CLO, among others, includes CQO and semidefinite optimization (SDO). In all these cases CLO problems can be solved efficiently by interior point methods (IPMs).

LO, QO, and SOCO formulations are presented below and their parametric (multiobjective) counterparts are discussed in Sect. 5.3. We leave parametric SDO outside of this chapter, even though there are some results available for this class of problems [25]. However, according to our best knowledge, there are no algorithms for parametric SDO that are implementation-ready and can be used in practice.

Linear optimization problems are formulated as:

$$
\begin{array}{ll}
\textit{Primal problem} & \textit{Dual problem} \\
(LP) \quad \min_{x}\{c^T x : Ax = b, x \geq 0\} & (LD) \quad \max_{y,s}\{b^T y : A^T y + s = c, s \geq 0\},
\end{array}
\tag{5.4}
$$

where $A \in \mathbb{R}^{m \times n}$, rank$(A) = m$, $c \in \mathbb{R}^n$, $b \in \mathbb{R}^m$, $x,s \in \mathbb{R}^n$, $y \in \mathbb{R}^m$.

Convex quadratic optimization problems contain a convex quadratic term in the objective function:

$$
\begin{array}{ll}
\textit{Primal problem} & \textit{Dual problem} \\
\quad \min_{x} c^T x + \tfrac{1}{2} x^T Q x & \quad \max_{x,y,s} b^T y - \tfrac{1}{2} x^T Q x \\
(QP) \quad \text{s.t. } Ax = b & (QD) \quad \text{s.t. } A^T y + s - Q x = c \\
\qquad\quad x \geq 0 & \qquad\qquad\quad s \geq 0,
\end{array}
\tag{5.5}
$$

where $Q \in \mathbb{R}^{n \times n}$ is a symmetric positive semidefinite matrix.

In *SOCO* problems the variables are restricted to lie in the Lorentz cone leading to the following formulation:

$$
\begin{array}{ll}
\textit{Primal problem} & \textit{Dual problem} \\
\min_{x} c^T x & \max_{y,s} b^T y \\
(SOCP) \quad \text{s.t.} \ Ax = b & (SOCD) \quad \text{s.t.} \ A^T y + s = c \\
\quad x_1^i \ge \|x_{2:n_i}^i\|, \ i = 1, \dots, I & \quad s_1^i \ge \|s_{2:n_i}^i\|, \ i = 1, \dots, I,
\end{array}
\tag{5.6}
$$

where $x = (x_1^1, x_2^1, \dots, x_{n_1}^1, x_1^2, x_2^2, \dots, x_{n_2}^2, \dots, x_1^I, x_2^I, \dots, x_{n_I}^I)^T \in \mathbb{R}^n$ and $s = (s_1^1, s_2^1, \dots, s_{n_1}^1, s_1^2, s_2^2, \dots, s_{n_2}^2, \dots, s_1^I, s_2^I, \dots, s_{n_I}^I)^T \in \mathbb{R}^n$ with $n = \sum_{i=1}^I n_i$. Second-order cone constraints of the type $x_1^i \ge \|x_{2:n_i}^i\|$ are often written as $(x_1^i, \bar{x}^i) \in \mathcal{K}_q^i$, where $\bar{x}^i = x_{2:n_i}^i$, or just $(x_1^i, \bar{x}^i) \ge_{\mathcal{K}_q^i} 0$.

As $(x_1^1, \dots, x_{n_1}^1)^T \in \mathcal{K}_q^1$, $(x_1^1, \dots, x_{n_2}^2)^T \in \mathcal{K}_q^2$, ..., $(x_1^I, \dots, x_{n_I}^I)^T \in \mathcal{K}_q^I$ and $\mathcal{K} = \mathcal{K}_q^1 \times \mathcal{K}_q^2 \times \dots \times \mathcal{K}_q^I$, we can also rewrite problem (5.6) in its shorter form (5.1). In the remainder of the chapter, cone $\mathcal{K}$ denotes the quadratic cone (direct product of linear cones $\mathcal{K}_\ell$ and quadratic cones $\mathcal{K}_q$), unless otherwise specified.

In addition to LO and QO problems, SOCO includes QCQO. Details about the QCQO problem formulation and its transformation to SOCO formulation can be found in [36].

LO, QO, and SOCO problems presented in this section are single-objective convex optimization problems. Most of the real-life optimization problems are multiobjective in their nature and in many cases those can be formulated as multiobjective LO, QO, or SOCO problems. Theoretical background and solution techniques for multiobjective optimization are discussed in Sect. 5.2. In that section we also highlight the relationships between multiobjective optimization and parametric optimization that is used to solve such problems. Parametric optimization algorithms for LO, QO, and SOCO optimization problems are the subject of Sect. 5.3. Extensions to other classes of optimization problems, e.g., convex nonlinear optimization, are briefly mentioned. Finally, we present financial applications of multiobjective optimization and numerically solve three examples in Sect. 5.4.

While solving multiobjective optimization problems via parametric optimization techniques as described in Sect. 5.2 can be done for any number of objective functions in a multiobjective optimization problem, we outline parametric optimization algorithms and provide examples in Sects. 5.3 and 5.4 for bi- and tri-objective optimization problems. Going beyond uni- and bi-parametric optimization (when the number of objective functions is more than three) is challenging from both computational and visualization point of view and is mostly beyond the scope of this chapter.

5.2 Multiobjective and Parametric Optimization

Let x be an n-dimensional vector of *decision variables*. The multiobjective optimization problem, where the goal is to optimize a number of objectives simultaneously, is formulated as:

$$\min \{f_1(x), f_2(x), \ldots, f_k(x)\}$$
$$\text{s.t. } x \in \Omega, \tag{5.7}$$

where $f_i : \mathbb{R}^n \to \mathbb{R}$, $i = 1, \ldots, k$ are (possibly) conflicting objectives and $\Omega \subseteq \mathbb{R}^n$ is a feasible region. Each of the functions f_i represents an attribute or a *decision criterion* that serves the base for the decision-making process.

Multiobjective optimization is a subclass of *vector optimization*, where the vector-valued objective function $f_0 = \{f_1(x), f_2(x), \ldots, f_k(x)\}$ is optimized with respect to a proper convex cone $\mathcal{C}$ which defines preferences. When a vector optimization problem involves the cone $\mathcal{C} = \mathbb{R}_+$, it is known as a *multicriteria or multiobjective optimization* problem.

In this chapter we consider *convex multiobjective conic optimization* problems and most of the results hereafter are restricted to that problem class. Moreover, we also mention some of the results available for general multiobjective problems. Problem (5.7) is a *convex* multiobjective optimization problem if all the objective functions $f_1, \ldots, f_k$ are convex, and the feasible set Ω is convex as well. For example, it can be defined as $\Omega = \{x : g_j(x) \leq 0, \ h_j(x) = 0\}$, where the inequality constraint functions $g_j : \mathbb{R}^n \to \mathbb{R}$, $j = 1, \ldots, l$ are convex and the equality constraint functions $h_j : \mathbb{R}^n \to \mathbb{R}$, $j = 1, \ldots, m$ are affine. For LO, QO, and SOCO problems the set of constraints can be written as $\Omega = \{x : Ax = b, \ x \geq_{\mathcal{K}} 0\}$, where $\mathcal{K}$ is an appropriate convex cone and $Ax = b$ are the equality constraints with $A \in \mathbb{R}^{m \times n}$ and $b \in \mathbb{R}^m$. The set Ω is called the feasible region in the decision space or just the *decision space*.

Definition 5.1. A vector $x^* \in \Omega$ is Pareto optimal (or efficient solution) if there does not exist another $x \in \Omega$ such that $f_i(x) \leq f_i(x^*)$ for all $i = 1, \ldots, k$ and $f_j(x) < f_j(x^*)$ for at least one index j.

The set of all Pareto optimal (or efficient) solutions $x^* \in \Omega$ is called the Pareto optimal (efficient solution) set Ω_E.

As values of the objective functions are used for making decisions by the decision maker, it is conventional for multiobjective optimization to work in the space of the objective functions, which is called the *objective space*. By mapping the feasible region into the objective space, we get:

$$Z = \{z \in \mathbb{R}^k : z = ((f_1(x), f_2(x), \ldots, f_k(x))^T \ \forall x \in \Omega)\}.$$

The set Z is the set of objective values of feasible points, and it is referred to as the set of *achievable objective values*. Points in the achievable set Z can be ranked into efficient and non-efficient points (see Fig. 5.1) that lead to the definition of Pareto optimality.

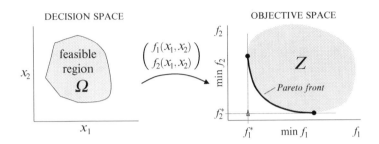

Fig. 5.1 Mapping the decision space into the objective space

Analogous definition of Pareto optimality can be stated for an objective vector $z^* \in Z$. Equivalently, z^* is Pareto optimal if the decision vector x^* corresponding to it is Pareto optimal [39].

Definition 5.2. For a given multiobjective problem (5.7) and Pareto optimal set Ω_E, the Pareto front is defined as:

$$Z_N = \{z^* = (f_1(x^*), \dots, f_k(x^*))^T \mid x^* \in \Omega_E\}.$$

A set Z_N of Pareto optimal (also called nondominated or efficient) solutions z^* forms the *Pareto efficient frontier* or *Pareto front*. The Pareto front, if $k = 2$, is also known as the *optimal trade-off curve* and for $k > 2$ it is called the *optimal trade-off surface* or the *Pareto efficient surface*.

Solution methods are designed to help the decision maker to identify and choose a point on the Pareto front. Identifying the whole frontier is computationally challenging, and often it cannot be performed in reasonable time. Solution methods for multiobjective optimization are divided into the following categories [39]:

- *A priori methods* are applied when the decision-maker's preferences are known a priori; those include the value function method, lexicographic ordering, and goal programming.
- *Iterative methods* guide the decision maker to identify a new Pareto point from an existing one (or existing multiple points); the process is stopped when the decision maker is satisfied with the actual efficient point.
- *A posteriori methods* are used to compute the Pareto front or some of its parts; those methods are based on the idea of scalarization, namely transforming the multiobjective optimization problem into a series of single-objective problems; a posteriori methods include *weighting methods*, the *ε-constrained method*, and related scalarization techniques.

Computing the Pareto front can be challenging as it does not possess known structure in most of the cases, and, consequently, discretization in the objective space is frequently used to compute it. The problem is that discretization is computationally costly in higher dimensions, and discretization is not guaranteed to produce all the (or desired) points on the Pareto front.

It turns out that for some classes of multiobjective optimization problems the structure of the efficient frontier can be identified. Those include multiobjective LO, QO, and SOCO optimization problems. For those classes of problems, the Pareto efficient frontier can be subdivided into pieces (subsets) that have specific properties. These properties allow the identification of each subset of the frontier. The piecewise structure of the Pareto front also provides additional information for the decision maker.

Before looking at the scalarization solution techniques for multiobjective optimization, which allow us to identify all nondominated (Pareto efficient) solutions, we need to introduce a number of concepts and some theoretical results.

Definition 5.3. An objective vector $z^* \in Z$ is weakly Pareto optimal if there does not exist another decision vector $z \in Z$ such that $z_i < z_i^*$ for all $i = 1, \ldots, k$.

The set of weakly Pareto efficient (nondominated) vectors is denoted by Z_{wN}. It follows that $Z_N \subseteq Z_{wN}$. When unbounded trade-offs between objectives are not allowed, Pareto optimal solutions are called *proper* [39]. The set of properly efficient vectors is denoted as Z_{pN}.

Both sets Z_{wN} (weak Pareto front) and Z_N (Pareto front) are connected if the functions f_i are convex and the set Ω satisfies one of the following properties [16]:

- Ω is a compact, convex set.
- Ω is a closed, convex set and $\forall z \in Z$, $\Omega(z) = \{x \in \Omega : f(x) \leq z\}$ is compact.

Let us denote by $\mathbb{R}_+^k = \{z \in \mathbb{R}^k : z \geq 0\}$ the nonnegative orthant of $\mathbb{R}^k$. Consider the set:

$$\mathcal{A} = Z + \mathbb{R}_+^k = \{z \in \mathbb{R}^k : f_i(x) \leq z_i, \ i = 1, \ldots, k, \ x \in \Omega\},$$

that consists of all values that are worse than or equal to some achievable objective value. While the set Z of achievable objective values need not be convex, the set $\mathcal{A}$ is convex, when the multiobjective problem is convex [8].

Definition 5.4. A set $Z \in \mathbb{R}^k$ is called $\mathbb{R}_+^k$-convex if $Z + \mathbb{R}_+^k$ is convex.

A point $x \in \mathcal{C}$ is a minimal element with respect to componentwise inequality induced by $\mathbb{R}_+^k$ if and only if $(x - \mathbb{R}_+^k) \cap \mathcal{C} = x$. The minimal elements of $\mathcal{A}$ are exactly the same as the minimal elements of the set Z. This also means that any hyperplane tangent to the Pareto efficient surface is a supporting hyperplane — the Pareto front is on one side of the hyperplane [14]. It follows that the Pareto front must belong to the boundary of Z [14].

Proposition 5.1. $Z_N = \left(Z + \mathbb{R}_+^k\right)_N \subset \mathrm{bd}(Z)$.

When talking about convex multiobjective optimization problems, it is useful to think of the Pareto front as a function, and not as a set. Under assumptions about convexity of the functions f_i and the set Ω for bi-objective optimization problems ($k = 2$), the (weakly) Pareto front is a convex function [45]. Unfortunately, when $k > 2$ it is not the case even for linear multiobjective optimization problems.

Most a posteriori methods for solving multiobjective optimization problems are based on scalarization techniques. Let us consider the two most popular scalarization methods:

- Weighting method
- ε-Constrained method

5.2.1 Weighting Method

The idea of the weighting method is to assign weights to each objective function and optimize the weighted sum of the objectives. A multiobjective optimization problem can be solved with the use of the *weighting method* by optimizing single-objective problems of the type

$$\min \sum_{i=1}^{k} w_i f_i(x) \tag{5.8}$$
$$\text{s.t. } x \in \Omega,$$

where f_i is linear, convex quadratic or second-order conic function in our case, $\Omega \subseteq \mathbb{R}^n$ (convex), $w_i \in \mathbb{R}$ is the weight of the ith objective, $w_i \geq 0$, $\forall i = 1,\ldots,k$, and $\sum_{i=1}^{k} w_i = 1$. Weights w_i define the importance of each objectives. Due to the fact that each objectives can be measured in different units, the objectives may have different magnitudes. Consequently, for the weight to define the relative importance of objectives, all objectives should be normalized first. Some of the normalization methods are discussed in [26]. As we intend to compute the whole Pareto front, normalization is not required.

It is known that the weighting method produces weakly efficient solutions when $w_i \geq 0$ and efficient solutions if $w_i > 0$ for all $i = 1,\ldots,k$ [39]. For convex multiobjective optimization problems any Pareto optimal solution x^* can be found by the weighting method.

Let us denote by $\mathcal{S}(w,Z) = \{\hat{z} \in Z : \hat{z} = \mathrm{argmin}_{z \in Z} w^T z\}$ the set of optimal points of Z with respect to w. In addition, we define

$$\mathcal{S}(Z) = \bigcup_{w>0,\ \sum_{i=1}^{k} w_i=1} \mathcal{S}(w,Z), \qquad \mathcal{S}_0(Z) = \bigcup_{w\geq 0,\ \sum_{i=1}^{k} w_i=1} \mathcal{S}(w,Z).$$

As Z is $\mathbb{R}_+^k$-convex set in our case, we get [14]:

$$\mathcal{S}(Z) = Z_{pN} \subset Z_N \subset \mathcal{S}_0(Z) = Z_{wN}. \tag{5.9}$$

In addition, if $\hat{z}$ is the unique element of $\mathcal{S}(w,Z)$ for some $w \geq 0$, then $\hat{z} \in Z_N$ [14]. The last observation combined with (5.9) allows us to identify the whole (weak) Pareto front with the use of the weighting method.

5.2.2 ε-Constrained Method

For illustration purposes, we first consider a problem with two objective functions. Multiobjective optimization can be based on ranking the objective functions in descending order of importance. Each objective function is then minimized individually subject to a set of additional constraints that do not allow the values of each of the higher ranked functions to exceed a prescribed fraction of their optimal values obtained in the previous step. Suppose that f_2 has higher rank than f_1. We then solve

$$\min\{f_2(x) : x \in \Omega\},$$

to find the optimal objective value f_2^*. Next, we solve the problem

$$\min f_1(x)$$
$$\text{s.t. } f_2(x) \leq (1+\varepsilon)f_2^*,$$
$$x \in \Omega.$$

Intuitively, the hierarchical ranking method can be thought as saying "f_2 is more important than f_1 and we do not want to sacrifice more than ε percentage of the optimal value of f_2 to improve f_1."

Considering the general case of k objective functions and denoting the right-hand side term of the constraints on the objective functions' values by $\varepsilon_j = (1 + \varepsilon_j)f_j^*$, we get the following single-objective optimization problem, which is known as the ε-constrained method:

$$(MOC_\varepsilon) \qquad \begin{array}{l} \min f_\ell(x) \\ \text{s.t. } f_j(x) \leq \varepsilon_j, \ j = 1,\ldots,k, \ j \neq \ell \\ x \in \Omega. \end{array} \qquad (5.10)$$

Every solution x^* of the ε-constrained problem (5.10) is weakly Pareto optimal [39], so formulation (5.10) can be used to compute weak Pareto front Z_{wN}.

Let x^* solve (5.10) with $\varepsilon_j^* = f_j(x^*)$, $j \neq \ell$. Then x^* is *Pareto optimal* [11,15] if:

1. x^* solves (5.10) for every $\ell = 1,\ldots,k$.
2. x^* is the unique solution of (5.10).
3. Lin's conditions [34,35].

The third set of necessary and sufficient conditions for (strong) Pareto optimality of optimal solutions is described in [11] based on the results of Lin [34,35]. Let us define

$$\phi_\ell(\varepsilon) = \min\{f_\ell(x) : x \in \Omega, \ f_j(x) \leq \varepsilon_j \text{ for each } j \neq \ell\}.$$

The following theorem [11] establishes that x^* is Pareto optimal if the optimal value of (MOC_{ε^0}) is strictly greater than $f_\ell(x^*)$ for any $\varepsilon^0 \leq \varepsilon^*$.

Theorem 5.1. *Let x^* solve (5.10) with $\varepsilon_j^* = f_j(x^*)$, $j \neq \ell$. Then x^* is Pareto optimal solution if and only if $\phi_\ell(\varepsilon) > \phi_\ell(\varepsilon^*)$ for all ε such that $\varepsilon \leq \varepsilon^*$ and for each ε (5.10) has an optimal solution with finite optimal value.*

In many cases, conditions (2) and (3) can be verified to identify the Pareto front Z_N. For instance, the second condition holds when all the objective functions $f_j(x)$ are strictly convex. Condition (3) can be verified if function $\phi_\ell(\varepsilon)$ is computed by parametric optimization techniques, see Sect. 5.2.4.

5.2.3 Parametric Optimization

Optimization models typically contain two types of variables: those that can be changed, controlled, or influenced by the decision maker are called *parameters*, the remaining ones are the decision variables. Parameters arise because the input data of the problem are not accurate or is changing over time. The main interest of sensitivity analysis is to determine how known characteristics of the problem are changing by small perturbations of the data. However, if we go farther from the current parameter value, not only the current properties of the problem might not be valid, but also the problem may change significantly. Study of this situation is referred to as *parametric optimization*.

Let us consider a general convex parametric optimization problem

$$\phi(\lambda) = \min\{f(x,\lambda) \ : \ x \in \mathcal{M}(\lambda), \lambda \in \Lambda\}, \tag{5.11}$$

with a parameter vector λ and function $f(x,\lambda)$ that is convex in terms of x. Let $\phi(\lambda)$ be the optimal value function and $\psi(\lambda)$ is the optimal set map of (5.11), and for our purpose

$$\mathcal{M}(\lambda) = \{x \in X \ : \ g_i(x) \leq \lambda_i, i = 1, 2, \ldots, m\}, \tag{5.12}$$

where g_i are real-valued functions defined on X. Observe that both $\mathcal{M}(\lambda)$ and $\psi(\lambda)$ are two point-to-set maps.

In parametric optimization, in addition to the optimal value function $\phi(\lambda)$, the optimal solution set $\psi(\lambda)$ is considered as function of the parameter vector. Investigating their behavior is the aim of parametric optimization.

Among many properties defined for optimization models, the following two are important ones: *uniqueness of optimal solution* and *stability*. When there is a unique optimal solution for each parameter, the solution set map is a real-valued map (as opposed to set-valued maps). Uniqueness of optimal solution is often violated in practice, especially for large-scale problems. For LO and QO problems it leads to degeneracy of optimal solutions, causing difficulties and ambiguities in post-optimality analysis [44]. Having multiple optimal solutions converts the optimal solution map to a point-to-set map. Continuous feasible perturbation of the parameter imply continuous changes of the feasible set and the optimal value function. The optimal value function describes the behavior of the objective function regardless of the uniqueness of the optimal solution.

By stability we mean having some important invariant properties of the problem such as continuity of the optimal value function or its differentiability. Even though the notion of "*stability*" stands for many different properties, there are two main ones. In [3], it is used for describing the semicontinuity of the optimal value function as well as the study of upper Hausdorff semicontinuity of the optimal solution set, that is a set-valued function in general. This approach to stability for QO has also been studied in [33]. Most probably optimal value function has critical points. If a parameter value is not a critical point of the optimal value function (or far enough from a critical point), it assures the decision maker that this solution is stable.

There is another notion, "*critical region*", used in parametric optimization with numerous meanings. In the LO literature it may refer to the region for parameter values where the given optimal basis (might be degenerate and not unique) remains optimal [20]. In QO and LO [7] it might also refer to the region where the active constraint set remains the same [42]. It is worth mentioning that the existence of a strictly complementary optimal solution is guaranteed in LO. Strictly complementary optimal solutions define the optimal partition of the index set (Sect. 5.3.1). The optimal partition is unique for any LO problem, and it has direct relation to the behavior of the optimal value function. In parametric optimization, one may be interested in studying the behavior of the optimal solution set, while one also might want to investigate the behavior of the optimal value function. These differences lead to diverse results in the context of parametric optimization. Here, we consider stability and critical regions in the context of optimal partition (see Sect. 5.3.1).

Recall that every convex optimization problem is associated with a Lagrangian dual optimization problem (see, e.g., [8]). The optimal objective function value of the dual problem is a lower bound for the optimal value of the primal objective function value (weak duality property). When these two optimal values are equal (strong duality), i.e., when the duality gap is zero, then optimality is achieved. However, the strong duality property does not hold in general, but when the primal optimization problem is convex and the *Slater constraint qualification* holds, i.e., there exists a strictly feasible solution, strong duality is guaranteed. In some special cases like LO, QO, and SOCO, parameters in the right-hand side of the constraints are translated as parameters in the coefficients of the objective function of its dual problem. This helps us to unify the analysis.

In this chapter, we consider optimal partition invariancy for LO problems [44] and for QO problems [22] as a criterion for analyzing the behavior of the optimal value function (see Sect. 5.3.1). We also extend these definitions to SOCO problems.

5.2.4 Multiobjective Optimization via Parametric Optimization

By now, the reader may have understood that multiobjective optimization problems are closely related to and can be represented as parametric optimization problems. Consequently, we may use algorithms of parametric optimization to solve multiobjective optimization problems and to compute the Pareto fronts.

Before defining the relations between multiobjective optimization and parametric optimization more formally, we mention that multiobjective LO, QO, and, to some extent, SOCO problems can be efficiently solved by parametric optimization algorithms. Parametric optimization techniques exist for wider classes of problems, but computational complexity may prevent using those directly to identify efficient frontiers.

The main idea of this chapter is that we can solve multiobjective optimization problems using parametric optimization techniques. *A posteriori multiobjective optimization techniques* are based on parameterizing (scalarizing) the objective space and solving the resulting parametric problem. Consequently, parametric optimization algorithms can be utilized to solve multiobjective optimization problems.

Based on the weighting method (5.8) and choosing the vector of weights as $w = (\lambda_1,\ldots,\lambda_{k-1},1)^T \geq 0$, as w can be scaled by a positive constant, for the weighted objective function $\sum_i w_i f_i(x)$, we can formulate the parametric optimization problem with the λ_i parameters in the objective function as

$$\phi(\lambda_1,\ldots,\lambda_{k-1}) = \min \quad \lambda_1 f_1(x) + \ldots + \lambda_{k-1} f_{k-1}(x) + f_k(x)$$
$$\text{s.t.} \quad x \in \Omega, \tag{5.13}$$

for computing weakly Pareto optimal solutions, or $(\lambda_1,\ldots,\lambda_{k-1})^T > 0$ for computing Pareto optimal solutions. Formulation (5.13) is known as the *Lagrangian problem* [11] and possesses almost identical properties as the weighting problem (5.8).

Based on the ε-constrained method (5.10), we can present the following parametric problem:

$$\phi(\varepsilon_1,\ldots,\varepsilon_{k-1}) = \min \quad f_k(x)$$
$$\text{s.t} \quad f_i(x) \leq \varepsilon_i,\ i = 1,\ldots,k-1$$
$$x \in \Omega, \tag{5.14}$$

where $\varepsilon_1,\ldots,\varepsilon_{k-1}$ are parameters in the right-hand side of the constraints. In this case, the optimal value function $\phi(\varepsilon_1,\ldots,\varepsilon_{k-1})$ includes the Pareto front as a subset.

It is not hard to observe that the parametric problems (5.13) and (5.14) are equivalent to (5.8) and (5.10), respectively, but they are just written in the forms used in the parametric optimization literature. The relationships between those formulations and their properties are extensively studied in [11].

Algorithms and techniques developed for solving parametric optimization problems are described in Sect. 5.3. Note that the optimal value function $\phi(\varepsilon)$ of problem (5.14) is the boundary of the set $\mathcal{A}$ and the Pareto front is a subset of that boundary. These results are illustrated by examples in Sect. 5.4.

5.2.5 Multiobjective and Parametric Quadratic Optimization

Results described by now in Sect. 5.2 apply to general convex multiobjective optimization problems. In contrast, parametric optimization techniques discussed in this chapter apply to LO, QO and SOCO problems only. In this section we specialize the formulations presented in Sect. 5.2.4 to the parametric optimization problem classes described in Sect. 5.3.

We define the *multiobjective quadratic optimization* problem as a convex multiobjective problem with one convex quadratic objective function f_k and $k-1$ linear objectives $f_1, \ldots, f_{k-1}$ subject to linear constraints. For the multiobjective QO problem the weighted sum formulation (5.13) specializes to

$$
\begin{aligned}
\phi(\lambda_1, \ldots, \lambda_{k-1}) &= \quad \min \lambda_1 c_1^T x + \ldots + \lambda_{k-1} c_{k-1}^T x + \tfrac{1}{2} x^T Q x \\
&\text{s.t. } Ax = b \\
&\qquad x \geq 0,
\end{aligned} \tag{5.15}
$$

and the ε-constrained formulation (5.14) becomes

$$
\begin{aligned}
\phi(\varepsilon_1, \ldots, \varepsilon_{k-1}) &= \quad \min \tfrac{1}{2} x^T Q x \\
&\text{s.t. } c_i^T x \leq \varepsilon_i, \ i = 1, \ldots, k-1 \\
&\qquad Ax = b \\
&\qquad x \geq 0.
\end{aligned} \tag{5.16}
$$

Parametric QO formulations (5.15) and (5.16) can be solved with algorithms developed in Sect. 5.3. The uni-parametric case corresponds to an optimization problem with two objectives. A bi-parametric QO algorithm allows solving multiobjective QO problems with three objectives. Multiobjective problems with more than three objectives require multiparametric optimization techniques (see Sect. 5.3.4.1). Note that in formulations (5.15) and (5.16), parameters appear in the objective function and in the right-hand side of the constraints, respectively.

Multiobjective QO problems are historically solved by techniques that approximate the Pareto front [18,19]. An alternative approach is the parametric optimization discussed in this chapter. Examples of multiobjective QO problems appearing in finance are solved with parametric QO techniques in Sect. 5.4.

If we allow for more than one convex quadratic objective in the multiobjective optimization problem, formulations (5.15) and (5.16) become parametric QCQO. It happens due to the fact that now quadratic functions appear in the constraints as well. Parametric SOCO, which includes parametric QCQO problems, is a more general class of problems. Preliminary results for solving parametric SOCO problems are described in Sect. 5.3.3, and we also provide a multiobjective SOCO example in Sect. 5.4.3. Properties of multiobjective optimization problems with more than one convex quadratic objectives and linear constraints are discussed in [24].

As we learned in this section, multiobjective optimization problems can be formulated as parametric optimization problems. Some classes of multiobjective optimization problems that include linear and convex quadratic optimization problems can be efficiently solved using parametric optimization algorithms. Parametric optimization allows not only computing Pareto efficient frontiers (surfaces) but also identifying piecewise structures of those frontiers. Structural description of Pareto fronts gives functional form of each of its pieces and thus helps decision makers to make better decisions.

5.3 Solving Parametric Optimization Problems

Utilizing different approaches for solving multiobjective optimization problems via parametric optimization (see Sects. 5.2.4 and 5.2.5), we review methods and results of uni- and bi-parametric LO, QO, and SOCO problems in this section. Our methodology is based on the notion of optimal partition and we study the behavior of the optimal value function that contains the Pareto front. To save space, all proofs are omitted, and we refer the interested reader to [43] and the related publications listed there for more details.

5.3.1 Uni-Parametric Linear and Convex Quadratic Optimization

The primal and dual solutions sets of QO are denoted by $\mathcal{QP}$ and $\mathcal{QD}$, respectively. Observe that the QO problem reduces to a LO problem where $Q = 0$. For a primal–dual optimal solution (x^*, y^*, s^*), the complementarity property $x^{*T}s^* = 0$ holds, which is equivalent to $x_j^* s_j^* = 0$ for $j \in \{1, 2, \ldots, n\}$. A strictly complementary optimal solution further satisfies $x^* + s^* > 0$. The existence of this kind of optimal solutions is true only for LO, while there is no guarantee to have strictly complementary optimal solution for any other class of optimization problems. For QO and SOCO problems the existence of maximally complementary optimal solution is proved. A primal–dual maximally complementary optimal solution has the maximum number of positive components for both x and s. In this case the *optimal partition* can be uniquely identified by:

$$\mathcal{B} = \{ j : x_j > 0, \, x \in \mathcal{QP} \},$$
$$\mathcal{N} = \{ j : s_j > 0, \, (y, s) \in \mathcal{QD} \},$$
$$\mathcal{T} = \{ 1, 2, \ldots, n \} \setminus (\mathcal{B} \cup \mathcal{N}).$$

As mentioned previously, for LO the set $\mathcal{T}$ is always empty.

The *uni-parametric QO* problem, with the parameter in the right-hand side of the constraints, is defined as

$$(QP_\varepsilon) \quad \phi(\varepsilon) = \min \left\{ c^T x + \frac{1}{2} x^T Q x \, : \, Ax = b + \varepsilon \triangle b, \, x \geq 0 \right\}, \tag{5.17}$$

with its dual as

$$(QD_\varepsilon) \quad \max \{ (b + \varepsilon \triangle b)^T y - \frac{1}{2} x^T Q x \, : \, A^T y + s - Q x = c, \, s \geq 0, \, x \geq 0 \}, \tag{5.18}$$

where $\triangle b \in \mathbb{R}^m$ is the fixed perturbing vector. The corresponding sets of feasible solutions are denoted by $\mathcal{QP}_\varepsilon$ and $\mathcal{QD}_\varepsilon$, and the optimal solution sets as $\mathcal{QP}_\varepsilon^*$ and $\mathcal{QD}_\varepsilon^*$, respectively.

The optimal value function $\phi(\varepsilon)$ is a piecewise convex (linear when $Q = 0$) quadratic function over its domain. Points where the optimal partition changes are referred to as *transition points*. These are precisely the points where the representation of the optimal value function changes too. At these points, the optimal value functions fails to have first or second-order derivatives. As the number of tri-partitions of the index set is finite, there are a finite number of quadratic pieces of the optimal value function.

To find the representation of the optimal value function on the (invariancy) intervals between two consequent transition points, we only need to have primal–dual optimal solutions for two parameter values from that interval.

Theorem 5.2. *For two values $\varepsilon_1 < \varepsilon_2$ with identical optimal partition, the optimal partition is the same for all $\varepsilon \in [\varepsilon_1, \varepsilon_2]$. Moreover, if (x^1, y^1, s^1) and (x^2, y^2, s^2) are maximally complementary primal–dual optimal solutions at ε_1 and ε_2, then the optimal value function can be represented as*

$$\phi(\varepsilon) = \phi(0) + \varepsilon \triangle b^T y^1 + \frac{1}{2}\varepsilon^2 \triangle b^T (y^2 - y^1), \tag{5.19}$$

where $\phi(0)$ corresponds to the optimal value of the unperturbed problem at $\varepsilon = 0$ and $0 \in [\varepsilon_1, \varepsilon_2]$.

Proof. See [5, pp. 6-21]. □

Observe that for the LO case, when the optimal partition is fixed, then the dual optimal solution set is invariant and consequently, the coefficient of the square term is zero. Thus, the objective value function is linear. To find an invariancy interval one needs to solve two auxiliary LO problems when a primal–dual optimal solution is available at an arbitrary parameter value in this interval. Let $\sigma(v)$ denotes the support set of the vector v, i.e., the index set of nonzero components of the vector v.

Theorem 5.3. *Let $x^* \in \mathcal{QP}_\varepsilon^*$ and $(x^*, y^*, s^*) \in \mathcal{QD}_\varepsilon^*$ be given for arbitrary ε. Let $(\varepsilon_\ell, \varepsilon_u)$ denotes the invariancy interval that includes ε. Moreover, let $T = \{1, 2, \ldots, n\} \setminus (\sigma(x^*) \cup \sigma(s^*))$. Then*

$$\varepsilon_\ell = \min\{\varepsilon : Ax - \varepsilon \triangle b = b, x \geq 0, x^T s^* = 0, x_T = 0,$$
$$A^T y + s - Qx = c, s \geq 0, s^T x^* = 0, s_T = 0\},$$

$$\varepsilon_u = \max\{\varepsilon : Ax - \varepsilon \triangle b = b, x \geq 0, x^T s^* = 0, x_T = 0,$$
$$A^T y + s - Qx = c, s \geq 0, s^T x^* = 0, s_T = 0\}.$$

Proof. See [5, pp. 6-26]. □

Remark 5.1. In case of LO, Theorem 5.3 reduces to solving the following simpler problems:

$$\varepsilon_\ell = \min\{\varepsilon : Ax - \varepsilon \triangle b = b, x \geq 0, x^T s^* = 0\},$$
$$\varepsilon_u = \max\{\varepsilon : Ax - \varepsilon \triangle b = b, x \geq 0, x^T s^* = 0\}.$$

Observe that this interval might be unbounded from one side if the corresponding auxiliary LO problem is unbounded, and the interval is singleton (transition point), when the given optimal solution is maximally (strictly in LO case) complementary and then $\varepsilon_\ell = \varepsilon_u$.

Finding the left and right derivatives of the optimal value function at a transition point requires the solution of two LO problems, provided an arbitrary primal–dual optimal solution is given at this point.

Theorem 5.4. *With the notation of Theorem 5.3, let (x^*, y^*, s^*) be a given optimal solution pair at the specific transition point ε. The left and right derivatives of the optimal value function $\phi(\varepsilon)$ are given by the optimal values of the following optimization problems:*

$$\phi'_- = \min_{x,y,s}\{\triangle b^T y : Ax - \varepsilon\triangle b = b, x \geq 0, x^T s^* = 0, x_T = 0,$$
$$A^T y + s - Qx = c, s \geq 0, s^T x^* = 0, s_T = 0\},$$
$$\phi'_+ = \max_{x,y,s}\{\triangle b^T y : Ax - \varepsilon\triangle b = b, x \geq 0, x^T s^* = 0, x_T = 0,$$
$$A^T y + s - Qx = c, s \geq 0, s^T x^* = 0, s_T = 0\}.$$

Proof. See [5, pp. 6-27]. □

Remark 5.2. In case of LO, Theorem 5.4 reduces to solving the following simpler problems:

$$\phi'_- = \min\{\triangle b^T y : A^T y + s = c, s \geq 0, s^T x^* = 0\},$$
$$\phi'_+ = \max\{\triangle b^T y : A^T y + s = c, s \geq 0, s^T x^* = 0\}.$$

Observe that at a transition point ϕ'_- or ϕ'_+ might be infinite as one of the associated auxiliary LO problems might be unbounded. It is not possible to have both LO problems in Theorem 5.4 feasible and unbounded, and they are both bounded and feasible when the transition point is not an end point of the domain of the optimal value function. Moreover, if ε is not a transition point, then $\phi'_- = \phi'_+$ which in this case is the derivative of the optimal value function.

We refer the interested reader to [5] for the results when, in case of parametric QO, perturbation exists in the linear term of the objective function. Results for the case, when both the right-hand side and the linear term of the objective function are perturbed simultaneously with the same parameter, can be found in [22].

5.3.2 From Uni- to Bi-Parametric Optimization

Going from uni-parametric to bi-parametric optimization has different problem formulations. One formulation of bi-parametric optimization is to have one of the parameters in the right-hand side of the constraints and the second one in the objective function data. This point of view to bi-parametric optimization has been considered extensively in LO and QO [21, 22, 27].

Another formulation is considering these two parameters either both in the right-hand side of the constraints, or both in the objective function data. From now on, by bi-parametric optimization, we mean having both parameters in the objective or both in the right-hand side data. Analogous to the previous discussion, we omit the LO problem as a special case, and review the results for bi-parametric QO problem with parameters in the right-hand side of the constraints. The bi-parametric QO problem is defined as follows:

$$(QP_{\varepsilon,\lambda}) \quad \phi(\varepsilon,\lambda) = \min_x \left\{ c^T x + \frac{1}{2} x^T Q x \; : \; Ax = b + \varepsilon \triangle b^1 + \lambda \triangle b^2, \; x \geq 0 \right\},$$

and its dual as

$$(QD_{\varepsilon,\lambda}) \quad \max_{x,y,s} \left\{ (b + \varepsilon \triangle b^1 + \lambda \triangle b^2)^T y - \frac{1}{2} x^T Q x \; : \; A^T y + s - Q x = c, \; s \geq 0 \right\},$$

where $\triangle b^1, \triangle b^2 \in \mathbb{R}^m$ are the given perturbing vectors. The case when so-called critical regions are defined as regions where a given optimal basis remains optimal has been discussed thoroughly in [42]. As mentioned in the uni-parametric LO case, the invariancy region where the optimal partition remains invariant, includes possibly exponentially many critical regions. Moreover, on an invariancy region, the optimal value function has a specific representation, and two disjoint regions correspond to two different representations of the optimal value function. In multiobjective optimization, we are interested in the behavior of the optimal value function (which includes the Pareto front) instead of the optimal solutions set. Thus, we investigate a technique for identifying all invariancy regions and describing the behavior of the optimal value function for bi-parametric optimization problems.

Let the optimal partition be known for $(\varepsilon,\lambda) = (0,0)$. The invariancy region that includes the origin is a (possibly unbounded) polyhedral convex set. This region is denoted here as $\mathcal{IR}(\triangle b^1, \triangle b^2)$ and referred to as the actual invariancy region. To identify this region, we refine the algorithmic approach used in [30]. The results and methodology are analogous to the case of bi-parametric QO [23].

5.3.2.1 Detecting the Boundary of an Invariancy Region

Observe that an invariancy region might be a singleton or a line segment. We refer to these type of regions as *trivial* regions. In this section, we describe the tools to identify a nontrivial invariancy region. Recall that for $\varepsilon = \lambda$, the bi-parametric QO problem reduces to uni-parametric QO problem. This trivial observation suggests developing a method to convert the bi-parametric QO problem into a series of uni-parametric QO problems. We start with identifying points on the boundary of the invariancy region. To accomplish this, we select the lines passing through the origin as

$$\lambda = t\varepsilon. \tag{5.20}$$

For now, we assume that the slope t is positive. Substituting (5.20) into the problem $(QP_{\varepsilon,\lambda})$ converts it into the following uni-parametric QO problem:

$$\min\left\{ c^T x + \frac{1}{2}x^T Qx : Ax = b + \varepsilon\overline{\triangle b},\ x \geq 0 \right\}, \tag{5.21}$$

where $\overline{\triangle b} = \triangle b^1 + t\triangle b^2$. Now, we can solve two associated auxiliary LO problems from Theorem 5.3 to identify the range of variation for parameter ε when equation (5.20) holds. These two auxiliary LO problems are:

$$\varepsilon_\ell = \min_{\varepsilon,x,y,s}\left\{ \varepsilon : Ax - \varepsilon\overline{\triangle b} = b,\ x_\mathcal{B} \geq 0,\ x_{\mathcal{N}\cup\mathcal{T}} = 0, \right.$$
$$\left. A^T y + s - Qx - \lambda\overline{\triangle c} = c,\ s_\mathcal{N} \geq 0,\ s_{\mathcal{B}\cup\mathcal{T}} = 0 \right\}, \tag{5.22}$$

and

$$\varepsilon_u = \max_{\varepsilon,x,y,s}\left\{ \varepsilon : Ax - \varepsilon\overline{\triangle b} = b,\ x_\mathcal{B} \geq 0,\ x_{\mathcal{N}\cup\mathcal{T}} = 0, \right.$$
$$\left. A^T y + s - Qx - \lambda\overline{\triangle c} = c,\ s_\mathcal{N} \geq 0,\ s_{\mathcal{B}\cup\mathcal{T}} = 0 \right\}, \tag{5.23}$$

where $\pi = (\mathcal{B},\mathcal{N},\mathcal{T})$ is the optimal partition for $\varepsilon = \lambda = 0$.

Now, we can summarize the procedure for identifying all transition points (vertices) and transition lines (edges) in an invariancy region. Let us assume that we know an initial inner point of the invariancy region and one of the edges (Fig. 5.2a, b shows how to find an inner point of the region). We are going to "shoot" by solving subproblem (5.22) or (5.23) counterclockwise from the initial point to identify each edge (see Fig. 5.2c–f). As we already know one of the edges, we exclude all the angles $\alpha_{\exp}$ between the initial point and the two vertices v_1 and v_2 of the known edge from the candidate angles to shoot. So, we shoot in the angle $v_0 - v_2$ plus in the small angles β and 2β and identify the optimal partition in the two points we get. Here we find the invariancy region boundary between the vertex v_2 and the point we get when shooting in the angle 2β. If the optimal partition is the same for the points in the directions β and 2β, we compute the vertices of this new edge e_2 and verify if one of those corresponds to a vertex of the previously known edge e_1. If it is not the case, then bisection is used to identify the missing edges between e_1 and e_2. We continue in this manner until all edges of the invariancy region are identified.

5.3.2.2 Transition from an Invariancy Region to the Adjacent Invariancy Regions

The first step of the algorithm is to determine the bounding box for the values of ε. Due to the fact that ε is the parameter appearing in the constraints, the problem $(QP_{\varepsilon,\lambda})$ may become infeasible for large or small ε values. Determining the bounding box is done as in many computational geometry algorithms [13, 41]. To find the range of ε where the parametric problem $(QP_{\varepsilon,\lambda})$ is feasible, we solve the following problem starting from the initial point $(\lambda_0, \varepsilon_0)$:

$$\min\left\{ c^T x + \frac{1}{2}x^T Qx : Ax = b + \varepsilon\triangle b^1 + \lambda_0\triangle b^2,\ x \geq 0 \right\}. \tag{5.24}$$

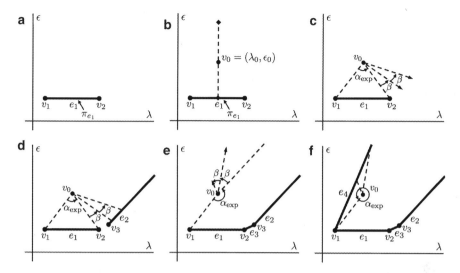

Fig. 5.2 Invariancy region exploration algorithm for bi-parametric QO

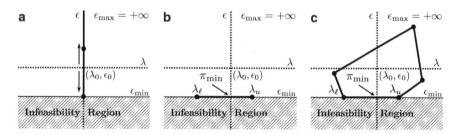

Fig. 5.3 The initialization of the bi-parametric QO algorithm

Solving problem (5.24) gives the values of $\varepsilon_{\min}$ and $\varepsilon_{\max}$ that (see Fig. 5.3a) are the lower and the upper feasibility bounds for the bi-parametric problem $(QP_{\varepsilon,\lambda})$. Observe that we may have either $\varepsilon_{\min} = -\infty$ or $\varepsilon_{\max} = +\infty$.

After identifying the feasibility bounds in the "$\varepsilon - \lambda$" plane, we choose $\varepsilon_{\min} \neq \infty$ or $\varepsilon_{\max} \neq \infty$. Let $\varepsilon = \varepsilon_{\min}$ and the optimal partition at the point $(\lambda_0, \varepsilon_{\min})$ is $\pi_{\min} = (\mathcal{B}_{\min}, \mathcal{N}_{\min}, \mathcal{T}_{\min})$. Then we can solve problems in Theorem 5.3 with the optimal partition $\pi = \pi_{\min}$ and $\lambda \triangle b^2$ replaced by $\varepsilon_{\min} \triangle b^2$ to identify the edge on the line $\varepsilon = \varepsilon_{\min}$ (see Fig. 5.3b). If the point $(\lambda_0, \varepsilon_{\min})$ is a singleton, we find the invariancy interval to the right from it. Now, we have an edge of one of the invariancy regions and we can get an initial inner point of that invariancy region selecting a point on the edge and utilizing Algorithm 6.3 from [5]. Using that initial inner point, we can identify the first nontrivial invariancy region including all of its edges and vertices as described in Sect. 5.3.2.1 (see Fig. 5.3c).

To enumerate all invariancy regions in the bounding box, we use concepts and tools [13, 41] from computational geometry. The algorithm that we are going to

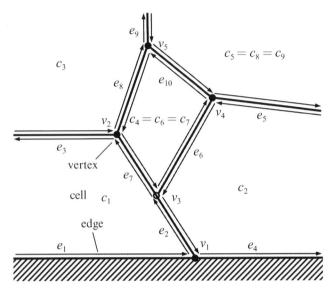

Fig. 5.4 Bi-parametric QO—computational geometry problem representation

present possess some similarities with polygon subdivision of the space and planar graphs. Our algorithm is essentially the subdivision of the bounding box into convex polyhedrons that can be unbounded.

The geometric objects involved in the given problem are vertices, edges, and cells (faces), see Fig. 5.4. Cells correspond to the nontrivial invariancy regions. Edges and vertices are trivial invariancy regions, each edge connects two vertices. It is important to notice that cells can be unbounded if the corresponding invariancy region is unbounded. That is why we need to extend the representation of the vertex to allow incorporating the information that the vertex can represent the virtual endpoint of the unbounded edge if the corresponding cell is unbounded. For instance, edge e_1 in Fig. 5.4 is unbounded, so in addition to its first endpoint v_1, we add another virtual endpoint being any point on the edge except v_1. Consequently, each vertex need to be represented not only by its coordinates (x, y), but also by the third coordinate z that indicates if it is a virtual vertex and the corresponding edge is unbounded. Another note to make is that the optimal partition may not be unique for each vertex or edge. First, at every virtual vertex, the optimal partition is the same as on the corresponding edge. Second, we may have situations when the optimal partition is the same on the incident edges and vertices if those are on the same line (edges e_2 and e_7 and vertex v_3 have the same optimal partition in Fig. 5.4).

To enumerate all invariancy regions we use two queues that store indices of the cells that are already investigated and to be processed. At the start of the algorithm, the first cell enters the to-be-processed queue and the queue of completed cells is empty (c_1 is entering the to-be-processed queue in Fig. 5.4). After that, we identify the cell c_1 including all faces and vertices starting from the known edge

Data: The CQO optimization problem and $\triangle b^1$, $\triangle b^2$
Result: Optimal partitions on all invariancy intervals, optimal value function
Initialization: compute bounding box in the "$\varepsilon - \lambda$" plane and compute inner point in one of the invariancy regions;
while *not all invariancy regions are enumerated* **do**
    run sub-algorithm to compute all edges and vertices of the current invariancy region;
    add all unexplored regions corresponding to each edge to the to-be-processed queue and move the current region to the queue of completed region indices;
    if *to-be-processed queue of the unexplored regions is not empty* **then**
        pull out the first region from the to-be-processed queue;
        compute an inner point of the new region;
    else
        return the data structure with all the invariancy regions, corresponding optimal partitions and optimal value function;
    end
end

Algorithm 1: Algorithm for enumerating all invariancy regions

e_1 and moving counterclockwise (note that the virtual vertices corresponding to the unbounded edges are not shown in Fig. 5.4). We continue in that manner until to-be-processed queue is empty and we have identified all the invariancy regions.

Algorithm 1 runs in linear time in the output size (the constant $C \cdot n$ is 3). But, by the nature of the parametric problem, the number of vertices, edges, and faces can be exponential in the input size. In our experiences worst case does not happen in practise very often though.

Remark 5.3. Similar to uni-parametric case, it is easy to verify that the optimal value function on an invariancy interval is a quadratic function in terms of the two parameters ε and λ, and it fails to have first or second directional derivatives passing the boundary of an invariancy region. Moreover, the optimal value function is a convex piecewise quadratic function on the "$\varepsilon - \lambda$" plane.

Remark 5.4. As we already mentioned, considering $Q = 0$ reduces a QO problem to a LO problem. Consequently, the outlined Algorithm 1 works for LO problems without any modifications. In bi-parametric LO case, the optimal value function is a piecewise linear function in two parameters ε and λ, and it fails to have directional derivative passing a transition line separating two adjacent invariancy regions.

5.3.3 Discussions on Parametric Second-Order Conic Optimization

Parametric SOCO is a natural extension of parametric analysis for LO and QO. As we point out in Sect. 5.2.5, parametric SOCO allows solving multiobjective quadratic optimization problems with more than one quadratic objective. The *optimal basis approach* to parametric optimization in LO cannot be directly generalized to parametric optimization in SOCO [46]. In contrast, it is promising to generalize the *optimal partition approach* of parametric LO and QO to SOCO. We describe ideas and preliminary results related to parametric SOCO in this section.

The standard form SOCO problem is defined in Sect. 5.1. Primal problem (*SOCP*) and dual problem (*SOCD*) are specified by Eq. (5.6). Before defining parametric SOCO formally, we describe the geometry of second-order (quadratic) cones. An extensive review of SOCO problems can be found in [2].

Unlike LO and QO, in SOCO we work with blocks of primal and dual variables, see the definition of (*SOCP*) and (*SOCD*) problems in Sect. 5.1. Those primal–dual blocks (x^i, s^i), $i = 1, \ldots, I$ of variables compose the decision vectors of SOCO problems $x = (x^1, \ldots, x^I)^T$ and $s = (s^1, \ldots, s^I)^T$, where $x^i, s^i \in \mathbb{R}^{n_i}$. We also refer to cone $\mathcal{K}$ as a second-order cone when it is a product cone $\mathcal{K} = \mathcal{K}_q^1 \times \ldots \times \mathcal{K}_q^I$, where $x^i \in \mathcal{K}_q^i$, $i = 1, \ldots, I$. As a linear cone $\mathcal{K}_\ell^i$ is a one-dimensional quadratic cone $\mathcal{K}_q^i$ $(x_1^i \geq 0)$, we treat linear variables as one-dimensional blocks. As before, $\mathcal{K}^*$ is the dual cone of $\mathcal{K}$.

The bi-parametric SOCO problem in primal and dual form is expressed as:

$$(SOCP_{\varepsilon,\lambda}) \qquad \begin{aligned} \phi(\varepsilon, \lambda) = \ & \min \ c^T x \\ & \text{s.t. } Ax = b + \varepsilon \triangle b^1 + \lambda \triangle b^2 \\ & \qquad x \in \mathcal{K}, \end{aligned} \tag{5.25}$$

and

$$(SOCD_{\varepsilon,\lambda}) \qquad \begin{aligned} & \max \ (b + \varepsilon \triangle b^1 + \lambda \triangle b^2)^T y \\ & \text{s.t. } A^T y + s = c \\ & \qquad s \in \mathcal{K}^*, \end{aligned} \tag{5.26}$$

where $A \in \mathbb{R}^{m \times n}$, $\text{rank}(A) = m$, $c \in \mathbb{R}^n$, $b \in \mathbb{R}^m$ are fixed data; $x, s \in \mathbb{R}^n$ and $y \in \mathbb{R}^m$ are unknown vectors; $\lambda, \varepsilon \in \mathbb{R}$ are the perturbation parameters. Note that constraints $x \in \mathcal{K}$ and $s \in \mathcal{K}^*$ are replaced by $x_1^i \geq \|x_{2:n_i}^i\|$ and $s_1^i \geq \|s_{2:n_i}^i\|$, $i = 1, \ldots, I$ for computational purposes.

Algebraic representation of the optimal partition for SOCO problems is required for computational purposes. It will allow identification of invariancy intervals for parametric SOCO problems.

Yildirim [47] has introduced an optimal partition concept for conic optimization. He took a geometric approach in defining the optimal partition while using an algebraic approach is necessary for algorithm design. Although the geometric approach has the advantage of being independent from the representation of the underlying optimization problem, it has some deficiencies. The major difficulty is extracting the optimal partition from a high-dimensional geometric object and, consequently, it is inconvenient for numerical calculations. In contrast, the algebraic approach is directly applicable for numerical implementation.

More recent study [6] provided the definition of the optimal partition that can be adapted to algebraic approach. We describe the details and compare the definitions of the optimal partition for SOCO in [47] (its algebraic form) and [6] in this section. Before defining the optimal partition for SOCO formally, we introduce the necessary concepts and notation. The interior and boundary of second-order cones are defined as follows.

Definition 5.5. The interior of second-order cone $\mathcal{K}_q \in \mathbb{R}^n$ is

$$\operatorname{int} \mathcal{K}_q = \{x \in \mathcal{K}_q : x_1 > \|x_{2:n}\|\}.$$

Definition 5.6. The boundary of second-order cone $\mathcal{K}_q \in \mathbb{R}^n$ without the origin 0 is

$$\operatorname{bd} \mathcal{K}_q = \{x \in \mathcal{K}_q : x_1 = \|x_{2:n}\|, \ x \neq 0\}.$$

Assuming strong duality, the *optimality conditions* for SOCO problems are:

$$
\begin{aligned}
Ax - b &= 0, \ x \in \mathcal{K}, \\
A^T y + s - c &= 0, \ s \in \mathcal{K}, \\
x \circ s &= 0,
\end{aligned}
$$

where the multiplication operation "$\circ$" is defined as $x \circ s = (x^1 \circ s^1, \dots, x^I \circ s^I)^T$ and $x^i \circ s^i = ((x^i)^T s^i, \ x_1^i s_2^i + s_1^i x_2^i, \ \dots, \ x_1^i s_{n_i}^i + s_1^i x_{n_i}^i)^T$.

Strict complementarity for SOCO problems [2] is defined as $x^i \circ s^i = 0$ and $x^i + s^i \in \operatorname{int}\mathcal{K}_q^i, i = 1, \dots, I$. IPMs for SOCO produce maximally complementary solutions that maximize the number of strictly complementary blocks i.

With respect to its cone $\mathcal{K}_q^i$ each block x^i can be in one of three states:

- block x^i is in the interior of $\mathcal{K}_q^i$:

$$\operatorname{int} \mathcal{K}_q^i = \{x^i \in \mathcal{K}_q^i : x_1^i > \|x_{2:n_i}^i\|\},$$

- block x^i is on the boundary of $\mathcal{K}_q^i$:

$$\operatorname{bd} \mathcal{K}_q^i = \{x^i \in \mathcal{K}_q^i : x_1^i = \|x_{2:n_i}^i\| \text{ and } x^i \neq 0\},$$

- block x^i equals 0:

$$x^i = 0.$$

The same results are valid for the dual blocks of variables $s^i \in (\mathcal{K}_q^i)^*$. As second-order cones are self-dual $\mathcal{K} = \mathcal{K}^*$, we are going to denote both primal and dual cones by $\mathcal{K}$ in the remainder of this chapter.

The optimal partition for SOCO has four sets, so it is a four-partition $\pi = (\mathcal{B}, \mathcal{N}, \mathcal{R}, \mathcal{T})$ of the index set $\{1, 2, \dots, I\}$. The four subsets are defined in [6] as:

$$\mathcal{B} = \{i : x_1^i > \|x_{2:n_i}^i\| \ (x^i \in \operatorname{int}\mathcal{K}_q^i) \text{ for a primal optimal solution } x\},$$

Table 5.1 Optimal partition for SOCO

	x^i		
s^i	0	bd $\mathcal{K}_q^i$	int $\mathcal{K}_q^i$
0	$i \in \mathcal{T}$	$i \in \mathcal{T}$	$i \in \mathcal{B}$
bd $\mathcal{K}_q^i$	$i \in \mathcal{T}$	$i \in \mathcal{R}$	$\times$
int $\mathcal{K}_q^i$	$i \in \mathcal{N}$	$\times$	$\times$

$$\mathcal{N} = \{\, i : s_1^i > \|s_{2:n_i}^i\| \ (s^i \in \operatorname{int}\mathcal{K}_q^i) \text{ for a dual optimal solution } (y,s)\,\},$$

$$\mathcal{R} = \{\, i : x^i \neq 0 \neq s^i \ (x^i \in \operatorname{bd}\mathcal{K}_q^i \text{ and } s^i \in \operatorname{bd}\mathcal{K}_q^i)$$

$$\text{for a primal–dual optimal solution } (x,y,s)\,\},$$

$$\mathcal{T} = \{\, i : x^i = s^i = 0, \text{ or } s^i = 0 \text{ and } s^i \in \operatorname{bd}\mathcal{K}_q^i, \text{ or } s^i = 0 \text{ and } x^i \in \operatorname{bd}\mathcal{K}_i^q$$

$$\text{for a primal–dual optimal solution } (x,y,s)\,\}.$$

Now we can state all possible configurations for primal–dual blocks of variables at optimality, those are summarized in Table 5.1 and serve as a basis for defining the optimal partition. Cases that are not geometrically possible, as those do not satisfy the optimality conditions, are shown as "$\times$" in Table 5.1.

For the set $\mathcal{R}$ of the optimal partition it holds that $x^i \neq 0 \neq s^i$, and those blocks x^i and s^i lie on the boundary of $\mathcal{K}$ (i.e., $x_1^i = \|x_{2:n_i}^i\| \neq 0$ and analogous relation holds for the dual). Let $(\bar{x}, \bar{y}, \bar{s})$ be a maximally complementary solution of problems $(SOCP)$ and $(SOCD)$ defined by (5.6), then as $x^i \circ s^i = 0$ we have

$$x^i \in \{\, \alpha\bar{x}^i : \alpha \geq 0\,\},$$

$$s^i \in \{\, \beta(\bar{x}_1^i, -\bar{x}_{2:n_i}^i) : \beta \geq 0\,\},$$

is equivalent to the primal and dual blocks belonging to orthogonal boundary rays of the cone $\mathcal{K}_q^i$.

We can replace the definition of the set $\mathcal{R}$ of the optimal partition by:

$$\mathcal{R}(\bar{x}) = \{\, (i,\bar{x}^i) : x^i \neq 0 \neq s^i, \ x^i \in \{\alpha\bar{x}^i : \alpha \geq 0\}, \ s^i \in \{\beta(\bar{x}_1^i, -\bar{x}_{2:n_i}^i) : \beta \geq 0\}$$
$$\text{for a primal–dual optimal solution } (x,y,s)\,\}.$$

Based on the results of Yildirim [47], we can alternatively define the optimal partition in algebraic form as $\pi_r = (\mathcal{B}, \mathcal{N}, \mathcal{R}(\bar{x}), \mathcal{T})$. The difference from the definition of π is that for primal–dual boundary blocks, it holds that $x^i \in$ a specific boundary ray of $\mathcal{K}_q^i$ and $s^i \in$ the orthogonal boundary ray of $\mathcal{K}_q^i$, instead of $x^i \in \operatorname{bd}\mathcal{K}_q^i$ and $s^i \in \operatorname{bd}\mathcal{K}_q^i$.

Comparing the two definitions of the optimal partition, π and π_r, it is worth to mention a couple of differences. When the optimal partition is defined as π, it partitions the index set $\{1, 2, \ldots, I\}$ of the blocks of variables. Consequently, it directly extends the definition of the optimal partition for QO (see Sect. 5.3.1) by adding the additional set $\mathcal{R}$ that corresponds to primal–dual optimal solutions being on the boundary of the cone, i.e., the case that does not exist for QO. In contrast, when the optimal partition is defined as π_r, it partitions not only the index set $\{1, 2, \ldots, I\}$, but also the space, as the set $\mathcal{R}(\bar{x})$ includes both indices of the blocks i and vectors $\bar{x}^i$ that define specific boundary rays. Definition of the optimal partition π_r is similar to the definition of the optimal partition for SDO in [25], which partitions the space and not the index set. Note that the real meaning of the partition set $\mathcal{R}(\bar{x})$ is that primal and dual vectors should be on the boundary of the cone and belong to a specific ray on that boundary. If the optimal solution stays on the boundary, but moves to another boundary ray when the problem $(SOCP)$ is perturbed, the optimal partition π_r changes, while π remains invariant.

Lets us consider the bi-parametric SOCO problem (5.25)–(5.26). We assume that the unperturbed problem $(SOCP_{0,0})$, where $\lambda = \varepsilon = 0$, has nonempty primal and dual optimal solution sets and strong duality holds for it, i.e., the duality gap is zero. For now, we use the definition π_r of the optimal partition.

Similar to parametric QO in Sect. 5.3.2, we can transform the bi-parametric SOCO problem into a series of uni-parametric problems. For simplicity, let us assign $\lambda = \varepsilon$. Moreover, let (x^*, y^*, s^*) be a maximally complementary optimal solution for $\varepsilon = 0$ with the optimal partition $\pi_r = (\mathcal{B}, \mathcal{N}, \mathcal{R}(x^*), \mathcal{T})$, the endpoints of the invariancy interval containing ε can be computed as:

$$\varepsilon_\ell = \min_{\varepsilon, x, y, s, \alpha, \beta} \left\{ \varepsilon : Ax - \varepsilon(\triangle b^1 + \triangle b^2) = b, \; x_{\mathcal{B} \cup \mathcal{T}} \in \mathcal{K}_{\mathcal{B} \cup \mathcal{T}}, \; x_{\mathcal{N}} = 0, \; x_{\mathcal{R}} = \alpha x_{\mathcal{R}}^*, \right.$$
$$\left. \alpha \geq 0, A^T y + s = c, \; s_{\mathcal{N} \cup \mathcal{T}} \in \mathcal{K}_{\mathcal{N} \cup \mathcal{T}}, \; s_{\mathcal{B}} = 0, \; s_{\mathcal{R}} = \beta s_{\mathcal{R}}^*, \; \beta \geq 0 \right\},$$

$$\varepsilon_u = \max_{\varepsilon, x, y, s, \alpha, \beta} \left\{ \varepsilon : Ax - \varepsilon(\triangle b^1 + \triangle b^2) = b, \; x_{\mathcal{B} \cup \mathcal{T}} \in \mathcal{K}_{\mathcal{B} \cup \mathcal{T}}, \; x_{\mathcal{N}} = 0, \; x_{\mathcal{R}} = \alpha x_{\mathcal{R}}^*, \right.$$
$$\left. \alpha \geq 0, A^T y + s = c, \; s_{\mathcal{N} \cup \mathcal{T}} \in \mathcal{K}_{\mathcal{N} \cup \mathcal{T}}, \; s_{\mathcal{B}} = 0, \; s_{\mathcal{R}} = \beta s_{\mathcal{R}}^*, \; \beta \geq 0 \right\},$$

where $\mathcal{K}_{\mathcal{B} \cup \mathcal{T}}$ is the Cartesian product of the cones $\mathcal{K}_q^i$ such that $i \in \mathcal{B} \cup \mathcal{T}$, $\mathcal{K}_{\mathcal{N} \cup \mathcal{T}}$ is defined analogously. Proof of this result for computing ε_ℓ and ε_u can be found in Theorem 4.1 in [47]. Alternatively, the constraints of the problems above can be completely rewritten in terms of the solution set instead of the index set, i.e., constraints $\{x_{\mathcal{B} \cup \mathcal{T}} \in \mathcal{K}, x_{\mathcal{N}} = 0, x_{\mathcal{R}} = \alpha x_{\mathcal{R}}^*, \alpha \geq 0\}$ can be written as $\{x \in \mathcal{K}, x \circ s^* = 0\}$.

The optimization problems for computing the endpoints ε_ℓ and ε_u of the current invariancy interval are SOCO optimization problems due to the fact that constraints of the form $x_{\mathcal{R}} = \alpha x_{\mathcal{R}}^*, \alpha \geq 0$ are linear (the invariancy interval can be a singleton, unlike in the QO case). In contrast, if we use the definition of the optimal partition π, constraints $x_{\mathcal{R}} \in \mathrm{bd}\,\mathcal{K}$ are nonlinear and are not second-order cone representable.

The results obtained by Yildirim [47] for the simultaneous perturbation case in conic optimization and by using the geometric definition of the optimal partition are directly linked to our findings. In his paper, Yildirim proved that the optimal value function is quadratic on the current invariancy interval. Although Yildirim's and our results are very interesting in the light of extending the parametric optimization techniques to SOCO problems, the obstacles, discussed in the remaining of this section, prevent direct mapping of them to algorithm design and implementation.

Unlike for parametric LO and QO problems, the optimal partition π_r for SOCO may change continuously, which poses difficulties for identifying all invariancy intervals for parametric SOCO. For the intervals of the parameter ε, where the optimal partition π_r is not changing continuously, the optimal value function is quadratic (see Proposition 5.1 in [47]). Another way to say it, for parametric SOCO we can have a continuum of changing transition points until we find an invariancy interval. In general, the optimal value function is piecewise-quadratic and it is quadratic on every invariancy interval. For the intervals, where the optimal partition changes continuously, we obtain the regions of nonlinearity of $\phi(\varepsilon)$ and there is no known way of describing $\phi(\varepsilon)$ completely on those intervals.

The intervals where the optimal partition π_r changes continuously, represent a curve on the boundary of the quadratic cone. Similarly, if the optimal partition is defined as π, the intervals with $\mathcal{R} \neq \emptyset$ represent a curve on the quadratic cone surface. Characterization of those curves and finding a computable description of them will allow identifying all invariancy intervals and computing the optimal value function. While those curves are conjectured to have hyperbolic shape, there are no results characterizing those curves that we are aware of. To get a computational algorithm for parametric SOCO, this characterization is a missing ingredient. Another remaining open problem is to find a rounding procedure for SOCO problems to identify exact optimal solutions.

Algorithms for computing the optimal value function $\phi(\varepsilon, \lambda)$ for parametric SOCO problems are subject of future research as there are no algorithms for parametric SOCO optimization. Invariancy regions corresponding to the definition of the optimal partition π are illustrated by an example in Sect. 5.4.3. That example also highlights the difficulties that arise during bi-parametric SOCO analysis.

5.3.4 Parametric Optimization: Extensions

5.3.4.1 Multiparametric Optimization

In this section we discuss how ideas from uni- and bi-parametric optimization in Sects. 5.3.1 and 5.3.2 extend to multiparametric case. Some multiparametric results exist for LO and QO from the optimal basis invariancy [42] and optimal partition invariancy [28]. Bi-parametric optimization algorithm from Sect. 5.3.2 can be extended to multiparametric case as well.

We would like to mention some differences of our algorithmic approach to parametric QO optimization in Sect. 5.3.2 and the algorithm described in [42] which is implemented in [32]. First, in our study we allow simultaneous perturbation in the right-hand side of the constraints and the linear term of the objective function with different parameters (see [23] for more details), while in [42] and related publications only perturbation in either the right-hand side or the linear term of the objective is considered. Second, in [42] the authors define a critical region as the region of parameters where active constraints remain active. As the result, an important precondition for analysis in [42] is the requirement for either making nondegeneracy assumption or exploiting special tools for handling degeneracy, while our algorithm does not require any nondegeneracy assumptions. Finally, the algorithm for parametric quadratic optimization described in [42] uses a different parameter space exploration strategy than ours. Their recursive algorithm identifies a first critical (invariancy) region and after that reverses the defining hyperplanes one by one in a systematic process to get a subdivision of the complement set. The regions in the subdivision are explored recursively. As the result, each critical (invariancy) region can be split among many regions and, consequently, all the parts has to be detected. Thus, each of the potentially exponential number of invariancy regions may be split among exponential number of regions, which makes their algorithm computationally expensive.

5.3.4.2 Nonlinear Parametric Optimization

Let us consider convex nonlinear parametric problem (5.11). When continuity of the functions $g_i(x, \lambda)$ for all (x, λ) and the convexity of these functions on $\mathbb{R}^n$ for all $\lambda \in \Lambda$ are added to the solution set (5.12), one can derive stronger results (see Sect. 3.2 in [3]).

This is the case we encounter in multiparametric LO and QO problems in some sense. With these assumptions, $\mathcal{M}$ is Hausdorff upper semi-continuous at λ_0 if $M(\lambda_0)$ is bounded and an $x_0 \in X$ exists such that $g(x_0) < \lambda_0$ ($g(x) = (g_1(x), \dots, g_m(x))^T$ and λ_0 is an m-vector) (see Theorem 3.3.1 in [3]). It means that x_0 must be an interior point of the parametric solution set (5.12).

Moreover, for $X = \mathbb{R}^n$, if $\mathcal{M}(\lambda)$ is nonempty for all $\lambda \in \Lambda$ and $\mathcal{M}(\lambda_0)$ be affine subspace, then $\mathcal{M}$ is Hausdorff-continuous at λ_0 (see Theorem 3.3.3.2 in [3]). This is the case we have in multiparametric LO and QO problems when perturbation occurs in the right-hand side of constraints.

5.3.4.3 Computational and Numerical Challenges

In this section we briefly describe computational costs and numerical issues of solving parametric optimization problems in order to compute Pareto fronts.

The major disadvantage of the techniques that approximate (discretize) Pareto fronts is that we do not compute the structure of a front. In addition, convex combination of the frontier points is not even on the frontier. When approximated Pareto fronts are computed, solution points may be poorly distributed along a frontier, so adaptive methods maybe required to improve the distribution of Pareto optimal solutions (see, e.g., [31]).

In most cases computational cost for computing the frontiers with parametric optimization would exceed those for approximating the frontiers when only discrete solutions are computed (due to potentially large or even exponential number of invariancy regions). To decrease solution time when solving parametric optimization problems, we can employ a number of options:

- Solve a parametric optimization problem for the specified ranges of parameters λ_i or ε_i
- Specify initial values of parameters λ_i or ε_i and only compute current invariancy region (interval) or a certain number of neighboring regions

Those techniques are typically applied when computing the entire Pareto front is not required or it is too computationally intensive, e.g., for large-scale optimization problems.

Implementing parametric optimization into software packages remains one of the challenges due to, in particular, numerical issues. The implementation of the computational algorithm contains some complications that are worth to mention:

- Due to numerical errors the determination of the optimal partition and a maximally complementary optimal solution, or the determination of the support set for a given optimal solution is a troublesome task; in contrast with the theoretical results, the numerical solution produced by a QO solver may not allow to determine the optimal partition or support set with 100% reliability.
- Incorrectly determined optimal partition or support sets, as well as numerical difficulties, may prevent one of the auxiliary subproblems to be solved that can also be used as an indicator of incorrectly identified partitions.
- When identifying boundaries of invariancy regions, numerical sensitivities may prevent getting those correctly.

The interested reader can find more details about overcoming numerical issues and implementing parametric optimization algorithms in [43].

5.4   Multiobjective Optimization Applications and Examples

In this section we present examples of multiobjective optimization problems that can be formulated and solved via parametric optimization. Multiobjective optimization problems arise in many areas including engineering (maximize vehicle speed and maximize its safety), finance (maximize profit and minimize risk), environmental economics (maximize profit and minimize environmental impact), and health care (kill tumor and spare healthy tissues). Examples described in this chapter are financial optimization problems from the area of risk management and portfolio selection. For examples of multiobjective optimization problems appearing in engineering, we refer the reader to consult a vast literature on multidisciplinary design [1]. Health care applications include intensity modulated radiation therapy (IMRT) planning for cancer treatment among others. For instance, a multiobjective linear IMRT problem is studied in [12], where the authors formulate an optimization problem with three objectives and compute an approximation of Pareto efficient surface.

In portfolio optimization, the goal of investors is to obtain optimal returns in all market environments when risk is involved in every investment, borrowing, lending, and project planning activity. From the multicriteria analysis point of view, investors need to determine what fraction of their wealth to invest in which asset in order to maximize the total return and minimize the total risk of their portfolio. There are many risk measures used for quantitative evaluation of portfolio risk including variance, portfolio beta, value-at-risk (VaR), and conditional value-at-risk (CVaR) among others. In addition to risk measures, there are portfolio performance indicators: expected market return, expected credit loss, price earnings ratio, etc. The most famous portfolio management model that involves a risk-return tradeoff is the mean–variance portfolio optimization problem introduced by Markowitz [38]. The conflicting objectives in the Markowitz model are minimizing portfolio variance (risk) and maximizing expected return.

Multiobjective optimization is a natural tool for portfolio selection models as those involve minimizing one or several risk measures and maximizing a number of portfolio performance indicators. We describe three variants of multiobjective portfolio optimization problems and their corresponding parametric formulations:

1. Parametric LO (three linear objectives) in Sect. 5.4.1
2. Parametric QO (two linear objectives and one quadratic objective) in Sect. 5.4.2
3. Parametric SOCO (one linear objective, one quadratic objective and one second-order conic objective) in Sect. 5.4.3

5.4.1 Portfolio Selection with Multiple Linear Objectives

Here, we discuss the multiobjective portfolio selection problem, where the objective functions are linear. Those models are rooted in the capital asset pricing model (CAPM), where the risk measure of an asset or portfolio is given by its beta coefficient. CAPM is the equilibrium version of mean–variance theory. Due to measuring risk in terms of the beta coefficients, the objective function in the risk minimization problem is linear in portfolio weights. In [48] a decision tool for the selection of stock portfolios based on multiobjective LO is described. Linear objective functions of the problem are the return, price earnings ratio, volume of transactions, dividend yield, increase in profits and risk, which is expressed as the linear function of betas. The authors apply portfolio selection to a set of 52 stocks from the Athens Stock Exchange. We are going to briefly describe their model including objective functions and constraints and compute the Pareto front for three out of six objectives considered in [48]. Readers interested in full details of the formulation and data for the model may consult [48].

The decision variables in portfolio selection problems are the portfolio weights x_i, $i = 1, \ldots, N$, where N is the total number of assets available for investment. Portfolio weights define a proportion of total wealth (or total budget) invested in the corresponding stock. As a matter of convenience, sum of portfolio weights is normalized to one $\sum_{i=1}^{N} x_i = 1$. Denoting by r_i the expected market return of an asset i allows us to compute the portfolio market return as $r_P = \sum_{i=1}^{N} r_i x_i = r^T x$.

The beta coefficient β is a relative measure of systematic (nondiversifiable) risk, and it reflects the tendency of an asset to move with the market. As beta measures correlation with the market portfolio, it is calculated as $\beta_i = \frac{\text{Cov}(r_i, r_M)}{\text{Var}(r_M)}$, where r_i is the asset i return and r_M is the return of the market portfolio. If $\beta_i < 1$ then asset i has less systematic risk than the overall market and the opposite holds for $\beta_i > 1$. As a result, portfolio risk minimization can be expressed as the linear function of asset weights, namely $\{\min_x \beta^T x\}$.

Among the other six objectives that are considered in [48] is maximizing return $\{\max_x r^T x\}$ and minimizing price earnings ratio (P/E) $\{\min_x d^T x\}$. The price earnings ratio d_i for each stock is computed as share price in the stock market at time period t divided by earnings per share at period $t-1$. We could have computed the Pareto efficient surface for more than three objectives here, but we restrict our attention to only those three due to well-known difficulties with visualizing surfaces in more than three dimensions. Denoting the three objectives as $f_1 = -r^T x$, $f_2 = \beta^T x$ and $f_3 = d^T x$, we obtain the following parametric optimization problem:

$$\min\ -r^T x + \lambda_1 \beta^T x + \lambda_2 d^T x$$
$$\text{s.t.}\ \ x \in \Omega, \tag{5.27}$$

where Ω in [48] is the set of linear constraints that includes no-short-sales restriction $x \geq 0$; upper limits for the capital allocations $x_i \leq u_i, i = 1,\ldots,52$; specific preferences for some stocks of the form $x_j \geq l_i$; and the constraints on betas of the form that portion y of the capital will be allocated to stocks with $\beta \in \{\beta_1, \beta_2\}$ that are expressed as $\sum_{i \subseteq I} x_i = y$. Note that maximizing $r^T x$ is equivalent to minimizing $-r^T x$.

The parametric optimization problem that follows from the ε-constrained multi-objective formulation is the following:

$$\min_{x,t}\ -r^T x$$
$$\text{s.t.}\ \beta^T x + t_1 = \varepsilon_1$$
$$d^T x + t_2 = \varepsilon_2$$
$$\sum_i x_i = 1$$
$$\sum_{i \in I} x_i = 0.2$$
$$x \geq 0,\ t \geq 0, \tag{5.28}$$

where t_1, t_2 are the slack variables used to convert the linear inequality constrains into equality constraints and $\varepsilon = (\varepsilon_1, \varepsilon_2)^T$ is the vector of parameters. We have used a subset of the constraints $x \in \Omega$ from [48] for the ease of exposition and included the no short-sales constraint $x \geq 0$ and the constraint $\sum_{i \in I} x_i = 0.2$ stating

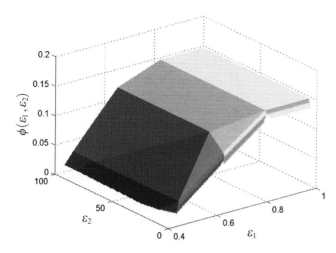

Fig. 5.5 The optimal value function for the parametric linear portfolio optimization problem

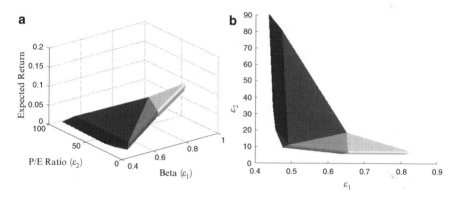

Fig. 5.6 The Pareto front for the multiobjective linear portfolio optimization problem (**a**) and the invariancy regions corresponding to it (**b**)

that 20% of capital is allocated to stocks with a beta coefficient less than 0.5. Formulation (5.28) is a parametric LO problem with two parameters in the right-hand side of the constraints.

The optimal value function for problem (5.28) is shown in Fig. 5.5. We can use the optimal partition for the variables t_1 and t_2 to determine the Pareto-efficient surface. For the invariancy regions corresponding to Pareto-efficient solutions, $t_1 \in \mathcal{N}$ and $t_2 \in \mathcal{N}$, meaning that those variables belong to the subset $\mathcal{N}$ of the optimal partition. The invariancy regions corresponding to the Pareto efficient solutions are shown in Fig. 5.6b and the Pareto front is depicted in Fig. 5.6a. The Pareto front is a piecewise linear function. The knowledge of invariancy intervals and optimal value function on those intervals gives us the structure of the Pareto front.

5.4.2 Mean–Variance Optimization with Market Risk and Transaction Cost

The Markowitz mean–variance model is commonly used in practice in the presence of market risk. From an optimization perspective, minimizing variance requires solving a QO problem. Denoting a vector of expected market returns by r as before and a variance–covariance matrix of returns by Q, the mean–variance portfolio optimization problem is formulated as a QO problem where the objectives are to maximize the expected portfolio return $\{\max_x r^T x\}$ and to minimize variance $\{\min_x x^T Q x\}$. The multiobjective optimization problem can be formulated as the weighted sum problem

$$\min_x \ -\lambda r^T x + \frac{1}{2} x^T Q x$$
$$\text{s.t. } x \in \Omega, \tag{5.29}$$

or as the ε-constrained problem

$$\min_x \ \frac{1}{2} x^T Q x$$
$$\text{s.t. } -r^T x \le \varepsilon,$$
$$x \in \Omega, \tag{5.30}$$

where Ω is the set of linear constraints on portfolio weights.

A portfolio may incur transaction cost associated with each trading. Denoting the linear transaction cost by ℓ_i, we add the third objective of minimizing the trading cost $\ell^T x$ of a portfolio to the mean–variance portfolio optimization problems (5.29)–(5.30).

We use a small portfolio optimization problem to illustrate the multiobjective optimization methodology. The problem data are presented in Tables 5.2 and 5.3. Table 5.2 shows expected market returns and per unit transaction cost for eight securities, as well as their weights in the initial portfolio x_0.

We put nonnegativity bounds $x \ge 0$ on the weights disallowing short-sales and optimize three objectives:

1. Minimize the variance of returns
2. Maximize expected market return
3. Minimize transaction cost

Moreover, we also need to add a constraint that makes the sum of the weights equal to one.

Thus, the multiobjective portfolio optimization problem looks like:

$$\min f_1(x) = -r^T x, \ f_2(x) = \ell^T x, \ f_3(x) = \frac{1}{2} x^T Q x$$
$$\text{s.t. } \sum_i x_i = 1,$$
$$x_i \ge 0 \ \forall i. \tag{5.31}$$

Table 5.2 Portfolio data for mean–variance optimization with market risk and transaction cost

Security	x_0	$r = \mathbb{E}(\text{market return})$	$\ell = (\text{transaction cost})$
1	0	0.095069	0.009830
2	0.44	0.091222	0.005527
3	0.18	0.140161	0.004001
4	0	0.050558	0.001988
5	0	0.079741	0.006252
6	0.18	0.054916	0.000099
7	0.13	0.119318	0.003759
8	0.07	0.115011	0.007334

We solve problem (5.31) as a parametric problem corresponding to the ε-constraint multiobjective formulation:

$$\min_{x,t} \frac{1}{2} x^T Q x$$
$$\text{s.t.} \ -r^T x + t_1 = \varepsilon_1$$
$$\ell^T x + t_2 = \varepsilon_2$$
$$\sum_i x_i = 1,$$
$$x \geq 0, \ t \geq 0, \tag{5.32}$$

where t_1, t_2 are the slack variables used to convert the linear inequality constrains into equality constraints and $\varepsilon = (\varepsilon_1, \varepsilon_2)^T$ is the vector of parameters.

The optimal value function for problem (5.32) is shown in Fig. 5.7 and the corresponding invariancy regions in Fig. 5.8a. We can utilize the optimal partition for the variables t_1 and t_2 to determine the Pareto efficient surface. For the invariancy regions corresponding to Pareto efficient solutions, $t_1 \neq \mathcal{B}$ and $t_2 \neq \mathcal{B}$, meaning that those variables do not belong to the subset $\mathcal{B}$ of the optimal partition. The invariancy regions corresponding to Pareto efficient solutions are shown in Fig. 5.8b and the Pareto front is depicted in Fig. 5.9.

Invariancy regions have a very intuitive interpretation for portfolio managers and financial analysts as inside each invariancy region the portfolio composition is fixed. By fixed composition we mean that the pool of assets included in the portfolio remains unchanged while asset weights can vary. For instance, on the invariancy region π_1 in Fig. 5.8b the optimal partition is $\mathcal{NNBBNBBN}$ which means that the portfolio is composed of securities 3, 4, 6, and 7. The functional form of the Pareto front on the invariancy region π_1 is $f_3 = 0.1 - 0.4f_1 - 23.7f_2 + 13.4f_1^2 + 11999.4f_2^2 - 621.9f_1f_2$.

Table 5.3 The return covariance matrix Q for mean–variance optimization with market risk and transaction cost

Security	1	2	3	4	5	6	7	8
1	0.002812	0.002705	−0.001199	0.000745	−0.000064	0.001035	−0.000336	0.000178
2	0.002705	0.015664	−0.003000	0.001761	−0.002282	0.007129	0.000596	−0.003398
3	−0.001199	−0.003000	0.008842	−0.000155	0.003912	0.001424	0.001183	−0.001710
4	0.000745	0.001761	−0.000155	0.002824	0.001043	0.003874	0.000225	−0.001521
5	−0.000064	−0.002282	0.003912	0.001043	0.007213	−0.001946	0.001722	0.001199
6	0.001035	0.007129	0.001424	0.003874	−0.001946	0.013193	0.001925	−0.004506
7	−0.000336	0.000596	0.001183	0.000225	0.001722	0.001925	0.002303	−0.000213
8	0.000178	−0.003398	−0.001710	−0.001521	0.001199	−0.004506	−0.000213	0.006288

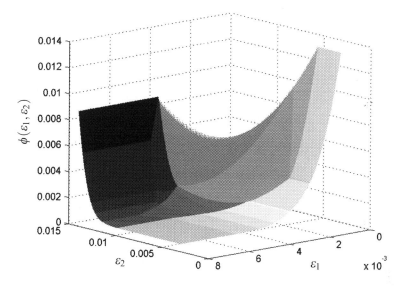

Fig. 5.7 The optimal value function for the mean–variance portfolio problem in the presence of transaction cost

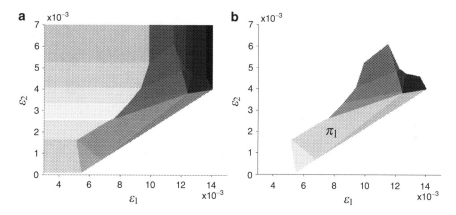

Fig. 5.8 Invariancy regions (**a**) and invariancy regions corresponding to the Pareto efficient solutions (**b**) for the mean–variance portfolio optimization problem with transaction cost

5.4.3 Robust Mean–Variance Optimization

One of the common criticisms of mean–variance optimization is its sensitivity to return estimates. As the consequence of that fact, small changes in the return estimates can result in big shifts of the portfolio weights x. One of the solutions to this problem is *robust optimization*, which incorporates uncertainties into the

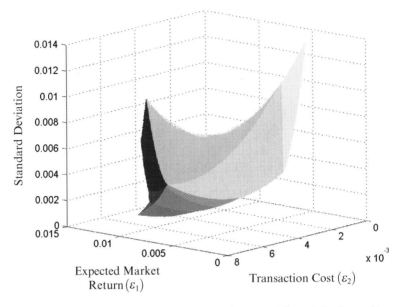

Fig. 5.9 The Pareto efficient surface for the mean–variance portfolio optimization problem with transaction cost

optimization problem. For a review of the robust optimization applied to portfolio management, we refer the reader to [17].

We consider a variant of robust portfolio selection problems proposed by Ceria and Stubbs [10]. In their model, instead of the uncertainty set being given in terms of bounds, they use ellipsoidal uncertainty sets. In [10] the authors assume that only r, the vector of estimated expected returns, is uncertain in the Markowitz model (5.29). In order to consider the worst case of problem (5.29), it was assumed that the vector of true expected returns r is normally distributed and lies in the ellipsoidal set

$$(r - \hat{r})^T \Theta^{-1} (r - \hat{r}) \leq \kappa^2,$$

with probability η, where $\hat{r}$ is an estimate of the expected return, Θ is covariance matrix of the estimates of expected returns, and $\kappa^2 = \chi_N^2(1 - \eta)$ with χ_N^2 being the inverse cumulative distribution function of the chi-squared distribution with N degrees of freedom.

Let $\hat{x}$ be the optimal portfolio on the estimated frontier for a given target risk level. Then, the worst case (maximal difference between the estimated expected return and the actual expected return) of the estimated expected returns with the given portfolio $\hat{x}$ can be formulated as:

$$\max_{\hat{r} - r} (\hat{r} - r)^T \hat{x}$$
$$\text{s.t.} \ (r - \hat{r})^T \Theta^{-1} (r - \hat{r}) \leq \kappa^2. \tag{5.33}$$

Table 5.4 Expected returns and standard deviations with correlations $= 20\%$ for robust mean–variance optimization, optimal weights for two portfolios

Security	r^1	r^2	σ	Security	Portfolio A	Portfolio B
Asset 1	7.15%	7.16%	20%	Asset 1	38.1%	84.3%
Asset 2	7.16%	7.15%	24%	Asset 2	69.1%	15.7%
Asset 3	7.00%	7.00%	28%	Asset 3	0.0%	0.0%

As derived in [10], by solving problem (5.33) we get that the optimal objective value $(\hat{r} - r)^T \hat{x}$ is $\kappa \|\Theta^{1/2}\hat{x}\|$. So, the true expected return of the portfolio can be expressed as $r^T \hat{x} = \hat{r}^T \hat{x} - \kappa \|\Theta^{1/2}\hat{x}\|$.

Now, problem (5.29) becomes a robust portfolio selection problem

$$\min_x \ -\lambda \hat{r}^T x + \frac{1}{2} x^T Q x + \kappa \|\Theta^{1/2}x\|,$$
$$\text{s.t. } x \in \Omega. \tag{5.34}$$

Problem (5.34) is SOCO problem; moreover, it is a parametric optimization problem. We solve an instance of problem (5.34) rewriting its formulation as:

$$\min \ -\hat{r}^T x + \kappa \|\Theta^{1/2}x\| + \lambda x^T Q x$$
$$\text{s.t. } \sum_{i=1}^{n} x_i = 1$$
$$x \geq 0, \tag{5.35}$$

where $\hat{r}$ is the vector of expected returns, Θ is the covariance matrix of estimated expected returns, Q is the covariance matrix of returns, κ is the estimation error aversion, and λ is the risk aversion.

Formulation (5.35) is a parametric SOCO problem with two parameters κ and λ. Preliminary results on solving parametric SOCO problems are discussed in Sect. 5.3.3. If we look at formulation (5.35) in the multiobjective sense, it is the problem of maximizing expected return, minimizing risk (variance of returns), and minimizing estimation error for the expected return. The problem formulation emphasizes the differences between the true, the estimated, and the actual Markowitz efficient frontiers [10].

To demonstrate the influence that sensitivities in the return estimates can potentially have on the portfolio selection, Ceria [9] presented a simple portfolio consisting of three assets. Table 5.4 shows expected returns for the two estimates and standard deviations for the assets. As Table 5.4 also shows, completely different portfolio weights can be obtained while optimizing the portfolio with expected return estimates r^1 and r^2. Taking r^1 as the estimate of the expected returns, we solve the multiobjective problem (5.35) to find all possible trade-offs between the three objectives – maximizing expected return, minimizing variance, and minimizing estimation error.

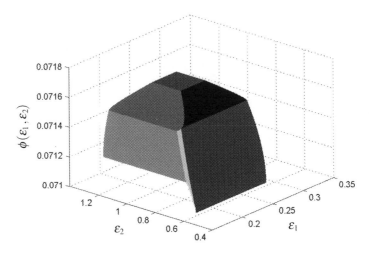

Fig. 5.10 The optimal value function for the robust mean–variance portfolio optimization problem

As $x^T Q x \leq \sigma_1^2$ $(Q = RR^T)$ and $\|\Theta^{1/2}x\| = \sqrt{x^T \Theta x} \leq \sigma_2$, we can rewrite problem (5.35) in the form:

$$
\begin{aligned}
\min \quad & -\hat{r}^T x + \lambda_1 u_0 + \lambda_2 v_0 \\
\text{s.t.} \quad & \textstyle\sum_{i=1}^n x_i = 1 \\
& x \geq 0 \\
& \Theta^{1/2}x - u = 0 \\
& R^T x - v = 0 \\
& (u_0, u) \in \mathcal{K}_q, \ (v_0, v) \in \mathcal{K}_q,
\end{aligned}
\tag{5.36}
$$

where parameters $\lambda_1 \geq 0$ and $\lambda_2 \geq 0$ and $\mathcal{K}_q$ is the second-order cone. Parametric problem (5.36) represents the weighting method for multiobjective optimization.

Formulating the parametric problem corresponding to the ε-constrained method for multiobjective optimization, we get:

$$
\begin{aligned}
\min \quad & -\hat{r}^T x \\
\text{s.t.} \quad & \textstyle\sum_{i=1}^n x_i = 1 \\
& x \geq 0 \\
& \Theta^{1/2}x - u = 0 \\
& R^T x - v = 0 \\
& u_0 = \varepsilon_1 \\
& v_0 = \varepsilon_2 \\
& (u_0, u) \in \mathcal{K}_q, \ (v_0, v) \in \mathcal{K}_q,
\end{aligned}
\tag{5.37}
$$

where parameters $\varepsilon_1 \geq 0$ and $\varepsilon_2 \geq 0$. Θ is the identity matrix in our example.

The optimal value function of the parametric SOCO formulation (5.37) with parameters $(\varepsilon_1, \varepsilon_2)$ in the constraints is shown in Fig. 5.10. The corresponding

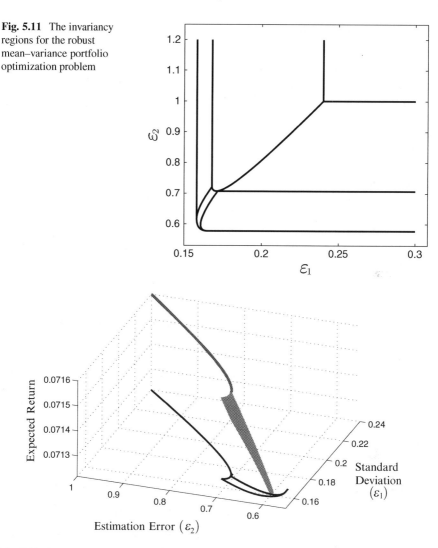

Fig. 5.11 The invariancy regions for the robust mean–variance portfolio optimization problem

Fig. 5.12 The Pareto efficient surface for the robust mean–variance portfolio optimization problem

invariancy regions are presented in Fig. 5.11. To identify the invariancy regions that correspond to Pareto efficient solutions, we need to restrict our attention to the regions where the second-order conic blocks u and v belong to the subsets $\mathcal{R}$ or $\mathcal{T}$ of the optimal partition. Those invariancy regions and the corresponding Pareto efficient surface are shown in Fig. 5.12.

5.5   Conclusions and Future Directions

In this chapter we considered techniques for solving multiobjective optimization problems and their parametric counterparts. By formulating and solving *multiobjective optimization* problems as *parametric optimization* problems, we bridged the gap between the two fields and unified the theory and practice of multiobjective and parametric optimization. Some classes of multiobjective optimization problems that include linear, convex quadratic, and potentially second-order conic optimization problems can be efficiently solved using parametric optimization algorithms. In particular, parametric optimization techniques described in this chapter give us a practical tool for solving multiobjective quadratic optimization problems. Parametric optimization allows not only computing Pareto fronts (efficient surfaces) but also identifying piecewise structure of those surfaces. Structural description of Pareto fronts gives functional form of each of its pieces and thus helps decision makers to make better decisions.

Even though some techniques exist for solving convex nonlinear parametric problems, those are not widely used. So, solving multiobjective convex nonlinear problems in practice is one of the hot research areas. If a multiobjective problem is non-convex (i.e., mixed integer), different approximations can be used allowing tracing Pareto efficient frontier with parametric optimization [42].

Integration of parametric optimization techniques that use optimal bases, optimal set invariancy, and optimal partition invariancy into a unified framework remains to be done. There are many publications that address different aspects of parametric optimization, but there is no study that puts those techniques together and describes how well those perform for different classes of optimization problems. Additional work has to be done on classifying multiobjective optimization problems for which the Pareto efficient frontier has identifiable structure.

Implementing parametric optimization into optimization software packages remains one of the challenges. Unfortunately, available software for parametric optimization is very limited. Commercial optimization packages such as CPLEX [29] and MOSEK [40] include basic sensitivity analysis for LO that is based on an optimal basis. MOSEK is the only package that provides optimal partition-based sensitivity analysis for LO as an experimental feature. As parametric optimization is the generalization of sensitivity analysis, techniques for identifying invariancy and stability regions have to be implemented on the top of sensitivity analysis available in those packages. Experimentation with active-set-based multiparametric optimization for LO and QO can be performed with MPT (multiparametric toolbox for MATLAB) [32] and this toolbox can be called from the YALMIP modeling environment [37].

Acknowledgements The authors' research was partially supported by the NSERC Discovery Grant #48923, the Canada Research Chair Program, and MITACS. The third author was also supported by a Start-up Grant of Lehigh University, and the Hungarian National Development Agency and the European Union within the frame of the project TAMOP 4.2.2-08/1-2008-0021 at the Széchenyi István University, entitled "Simulation and Optimization — basic research in numerical mathematics." We are grateful to Helmut Mausser, Alexander Kreinin, and Ian Iscoe from Algorithmics Incorporated, an IBM Company, for valuable discussions on the practical examples of multiobjective optimization problems in finance. We would like to thank Antoine Deza from McMaster University, Imre Pólik from SAS Institute and Yuri Zinchenko from University of Calgary for their helpful suggestions.

References

1. Alexandrov, N.M., Hussaini, M.Y. (eds.): Multidisciplinary Design Optimization: State of the Art. Proceedings in Applied Mathematics Series, No. 80. SIAM, Philadelphia (1997)
2. Alizadeh, F., Goldfarb, D.: Second-order cone programming. Math. Program. Ser. B **95**(1), 3–51 (2003)
3. Bank, B., Guddat, J., Klatte, D., Kummer, B., Tammer, K.: Non-Linear Parametric Optimization. Birkhäuser Verlag, Basel (1983)
4. Ben-Tal, A., Nemirovski, A.: Lectures on Modern Convex Optimization: Analysis, Algorithms, and Engineering Applications. MPS-SIAM Series on Optimization. MPS/SIAM, Philadephia (2001)
5. Berkelaar, A.B., Roos, C., Terlaky, T.: The optimal set and optimal partition approach to linear and quadratic programming. In: Gal, T., Greenberg, H.J. (eds.) Advances in Sensitivity Analysis and Parametric Programming, Chapter 6, pp. 6-1–6-44. Kluwer Academic Publishers, Boston (1997)
6. Bonnans, J.F., Ramírez C., H.: Perturbation analysis of second-order cone programming problems. Math. Program. Ser. B **104**(2–3), 205–227 (2005)
7. Borrelli, F., Bemporad, A., Morari, M.: Geometric algorithm for multiparametric linear programming. J. Optim. Theory Appl. **118**(3), 515–540 (2003)
8. Boyd, S.P., Vandenberghe, L.: Convex Optimization. Cambridge University Press, Cambridge (2004)
9. Ceria, S.: Robust portfolio construction. Presentation at Workshop on Mixed Integer Programming, University of Miami (June 5–8, 2006). http://coral.ie.lehigh.edu/mip-2006/talks/Ceria.ppt
10. Ceria, S., Stubbs, R.A.: Incorporating estimation errors into portfolio selection: robust portfolio construction. J. Asset Manage. **7**(1), 109–127 (2006)
11. Chankong, V., Haimes, Y.Y. (eds.): Multiobjective Decision Making: Theory and Methodology. Elsevier Science Publishing Co., New York (1983)
12. Craft, D.L., Halabi, T.F., Shih, H.A., Bortfeld, T.R.: Approximating convex Pareto surfaces in multiobjective radiotherapy planning. Med. Phys. **33**(9), 3399–3407 (2006)
13. de Berg, M., van Kreveld, M., Overmars, M., Schwarzkopf, O.: Computational Geometry: Algorithms and Applications, 2nd edn. Springer, Berlin (2000)
14. Ehrgott, M.: Multicriteria Optimization, 2nd edn. Springer, Berlin (2005)
15. Ehrgott, M.: Multiobjective optimization. AI Mag. **29**(4), 47–57 (2008)
16. Ehrgott, M., Wiecek, M.M.: Mutiobjective programming. In: Figueira, J., Greco, S., Ehrgott, M. (eds.) Multiple Criteria Decision Analysis: State of the Art Surveys. International Series in Operations Research & Management Science, vol. 78, Chapter 17, pp. 667–708. Springer, New York (2005)
17. Fabozzi, F.J., P.N. Kolm, D.A. Pachamanova, S.M. Focardi: Robust Portfolio Optimization and Management. Wiley, Hoboken (2007)
18. Fliege, J.: An efficient interior-point method for convex multicriteria optimization problems. Math. Oper. Res. **31**(4), 825–845 (2006)
19. Fliege, J., Heseler, A.: Constructing Approximations to the Efficient Set of Convex Quadratic Multiobjective Problems. Ergebnisberichte Angewandte Mathematik 211, Fachbereich Mathematik, Universitat Dortmund, Dortmund (2002)
20. Gal, T., Nedoma, J.: Multiparametric linear programming. Manage. Sci. **18**(7), 406–422 (1972)

21. Ghaffari-Hadigheh, A., Ghaffari-Hadigheh, H., Terlaky, T.: Bi-parametric optimal partition invariancy sensitivity analysis in linear optimization. Central Eur. J. Oper. Res. **16**(2), 215–238 (2008)
22. Ghaffari-Hadigheh, A., Romanko, O., Terlaky, T.: Sensitivity analysis in convex quadratic optimization: simultaneous perturbation of the objective and right-hand-side vectors. Algorithmic Oper. Res. **2**(2), 94–111 (2007)
23. Ghaffari-Hadigheh, A., Romanko, O., Terlaky, T.: Bi-parametric convex quadratic optimization. Optim. Methods Softw. **25**(2), 229–245 (2010)
24. Goh, C.J., Yang, X.Q.: Analytic efficient solution set for multi-criteria quadratic programs. Eur. J. Oper. Res. **92**, 166–181 (1996)
25. Goldfarb, D., Scheinberg, K.: On parametric semidefinite programming. Appl. Numer. Math. **29**(3), 361–377 (1999)
26. Grodzevich, O., Romanko, O.: Normalization and other topics in multi-objective optimization. In: Aruliah, D.A., Lewis, G.M. (eds.) Proceedings of the First Fields-MITACS Industrial Problems Workshop, pp. 89–101. Fields Institute for Research in Mathematical Sciences, Toronto (2006)
27. Guddat, J., Vasquez, F.G., Tammer, K., Wendler, K.: Multiobjective and stochastic optimization based on parametric optimization. In: Mathematical Research, vol. 26. Akademie-Verlag, Berlin (1985)
28. Hladík, M.: Multiparametric linear programming: support set and optimal partition invariancy. Eur. J. Oper. Res. **202**(1), 25–31 (2010)
29. ILOG Inc.: ILOG CPLEX User's Manual (2008). http://www.ilog.com. CPLEX 11.2
30. Kheirfam, B., Mirnia, K.: Quaternion parametric optimal partition invariancy sensitivity analysis in linear optimization. Adv. Model. Optim. **10**(1), 39–40 (2008)
31. Kim, I.Y., de Weck, O.L.: Adaptive weighted-sum method for bi-objective optimization: Pareto front generation. Struct. Multidisciplinary Optim. **29**(2), 149–158 (2005)
32. Kvasnica, M.: Real-Time Model Predictive Control via Multi-Parametric Programming: Theory and Tools. VDM Verlag, Saarbrücken (2009). http://control.ee.ethz.ch/~mpt/
33. Lee, G.M., Tam, N.N., Yen, N.D.: Continuity of the solution map in quadratic programs under linear perturbations. J. Optim. Theory Appl. **129**(3), 415–423 (2006)
34. Lin, J.G.: Three methods for determining Pareto-optimal solutions of multiple-objective problems. In: Ho, Y.C., Mitter, S.K. (eds.) Directions in Large-Scale Systems, pp. 117–138. Plenum, New York (1975)
35. Lin, J.G.: Proper inequality constraints and maximization of index vectors. J. Optim. Theory Appl. **21**(4), 505–521 (1977)
36. Lobo, M.S., Vandenberghe, L., Boyd, S.P., Lebret, H.: Applications of second-order cone programming. Linear Algebra Appl. **284**, 193–228 (1998)
37. Löfberg, J.: YALMIP: a toolbox for modeling and optimization in MATLAB. In: Proceedings of the CACSD Conference. Taipei, Taiwan (2004)
38. Markowitz, H.M.: Portfolio selection. J. Finance **7**(1), 77–91 (1952)
39. Miettinen, K.M.: Nonlinear Multiobjective Optimization. Kluwer Academic Publishers, Boston (1999)
40. MOSEK ApS: The MOSEK Optimization Tools Manual. Version 5.0 (2008). http://www.mosek.com. Revision 105
41. O'Rourke, J.: Computational Geometry in C, 2nd edn. Cambridge University Press, Cambridge (2001)
42. Pistikopoulos, E.N., Georgiadis, M.C., Dua, V. (eds.): Multi-Parametric Programming: Theory, Algorithms, and Applications, vol. 1. Wiley-VCH Verlag GmbH & Co. KGaA, Weinheim (2007)
43. Romanko, O.: Parametric and multiobjective optimization with applications in finance. Ph.D. Thesis, Department of Computing and Software, McMaster University, Hamilton (2010)
44. Roos, C., Terlaky, T., Vial, J.P.: Interior Point Methods for Linear Optimization. Springer Science, New York (2006)

45. Siem, A.Y.D., den Hertog, D., Hoffmann, A.L.: The effect of transformations on the approximation of univariate (convex) functions with applications to Pareto curves. Eur. J. Oper. Res. **189**(2), 347–362 (2008)
46. Yildirim, E.A.: An interior-point perspective on sensitivity analysis in linear programming and semidefinite programming. Ph.D. Thesis, School of Operations Research and Information Engineering, Cornell University, Ithaca (2001)
47. Yildirim, E.A.: Unifying optimal partition approach to sensitivity analysis in conic optimization. J. Optim. Theory Appl. **122**(2), 405–423 (2004)
48. Zopounidis, C., Despotis, D.K., Kamaratou, I.: Portfolio selection using the ADELAIS multiobjective linear programming system. Comput. Econ. **11**(3), 189–204 (1998)